T0391854

Writing the Brain

Writing the Brain

Material Minds and Literature, 1800–1880

STEFAN SCHÖBERLEIN

OXFORD
UNIVERSITY PRESS

Oxford University Press is a department of the University of Oxford. It furthers
the University's objective of excellence in research, scholarship, and education
by publishing worldwide. Oxford is a registered trade mark of Oxford University
Press in the UK and certain other countries.

Published in the United States of America by Oxford University Press
198 Madison Avenue, New York, NY 10016, United States of America.

© Oxford University Press 2023

All rights reserved. No part of this publication may be reproduced, stored in
a retrieval system, or transmitted, in any form or by any means, without the
prior permission in writing of Oxford University Press, or as expressly permitted
by law, by license, or under terms agreed with the appropriate reproduction
rights organization. Inquiries concerning reproduction outside the scope of the
above should be sent to the Rights Department, Oxford University Press, at the
address above.

You must not circulate this work in any other form
and you must impose this same condition on any acquirer.

CIP data is on file at the Library of Congress

ISBN 978–0–19–769368–1

DOI: 10.1093/oso/9780197693681.001.0001

Printed by Integrated Books International, United States of America

*Dedicated to Ed and Garrett,
in many ways the brains of this operation*

In the first half of the nineteenth century, a transformation took place in the natural sciences that, in its very nature as well as in the enormity of the revolution it triggered, may only be compared to that which had . . . come about in the late seventeenth century . . . and fulminated into the *Critique of Pure Reason.* . . .

Experiments on the body generated insights about the soul; in the very midst of the realm of physiology, we found ourselves face to face with the psyche: a knot was discovered that tightly bound these two modes of being to one another and, from that point onward, careful fingers could reach into the mysterious darkness of the mind. Thus, we confronted a moment of outrageous novelty in the history of science: the psyche, *pneuma*, that which is above and beyond mere things, the unfathomable itself, became flesh and took up residence within us.

—Gottfried Benn, 1910

Contents

Acknowledgments	ix
The First Century of the Brain: An Introduction	1
1. Nature's Mind and Mind's Nature: Romantic Cognition Between Harp and Atom	19
Harp Strings	19
Mind-Strings	29
Mind-Matter	37
Mind-Atoms	44
2. Split Brains, Doubled Minds: The Gothic's Bicameral Vision	59
The Sleepers	59
Dialoging the Self	67
Hemispheric Voices	75
Master-Minds	85
3. Skulls and Society: Reading the Mind as a Multi-Organ Entity	94
Brain Damage	94
Phrenology as Sociology	103
Phrenological Victorianism	107
Phrenological Americanism	113
Phrenology's Real	123
4. Cranial Reconstruction: Racialized Brains and the Psychometric Real	126
Uncommon Minds	126
Great Brains	131
A Cranial Case Study	136
Realism as Psychometry	143
5. Rattle-Brained: Insanity as Material Metacognition	157
You, Me, Brain	157
Metempsychosis as Metacognition	160
Insanity as Pop Culture	168
Psychosis as Metacognition	177

viii CONTENTS

6. The Telegraphed Brain: Wires as Proto-Neurons 195
    Thoughts on Wires 195
    Telegraphed Minds 208
    Introspective Brain-Machines 214

Afterword 227
References 233
Index 259

Acknowledgments

Like a phrenological skull's innumerable bumps and grooves, many people have left their impressions on this book that deserve to be acknowledged. *Writing the Brain* saw its beginnings as a doctoral dissertation at the University of Iowa (2018), and I would like to thank both the institution as well as the many people who had their hand in turning my cranial confusions into a somewhat coherent book: Loren Glass, Laura Rigal, John Durham Peters, Ed Folsom, and Garrett Stewart. I'm beyond grateful, in particular, to Ed and Garrett, to whom this book is dedicated. This "ill-form'd offspring of my feeble brain" would have never seen the light of day without their encouragement, feedback, and support during my years at Iowa and beyond. John Durham Peters, too, deserves a special "Danke," as does Blake Bronson-Bartlett, for always being willing to chat about media obscura and nineteenth-century oddities.

Thanks are also due to the many kind colleagues, students, and mentors that I had the pleasure to work with at the University of Iowa and Marshall University, as well as my many wonderful collaborators elsewhere. No matter how incomplete, this list has to include: Allison Carey, Kristen Lillvis, Hilton Córdoba, Victor Fet, Kateryna Schray (Данке), Chris White, Teri Reynoso, Kevin McMullen, Stephanie M. Blalock, Jason Stacy (master of the interface), Margee Miller, Naomi Greyser, Barbara Eckstein, Kathleen Diffley, Kembrew McLeod, Kenneth M. Price, Nicole Gray, Matt Cohen, Walter Grünzweig, Ted Widmer, and Zachary Turpin. I would also like to thank my new colleagues at Texas A&M University–Central Texas for welcoming me so warmly during the final weeks of this book's lengthy gestation.

The original spark for *Writing the Brain* came during conversations about Emily Dickinson's anatomical poetry with my wife, Arianna, who is truly a Renaissance person when it comes to all things "brain." *Grazie mille.* Our cats, Bernie and Hux, sat idly by during the writing of this book but deserve thanks nonetheless.

Of course, any piece of scholarship on a historical period is only as good as the sources its author could access. Without the untiring work of the Interlibrary Loan departments at the University of Iowa, Marshall University,

X ACKNOWLEDGMENTS

and Texas A&M University–Central Texas this book would have been much thinner. Additionally, digital archives including the Hathi Trust, Chronicling America, the Walt Whitman Archive, the Edgar Allan Poe Society website, the Charles Brockden Brown Electronic Archive, and the Internet Archive (among others) proved indispensable. May we never have to miss tools like these!

I would also like to gratefully acknowledge the journals that have given me permission to reprint elements of previously published articles, mainly my 2014 piece "Insane in the Membrane: Emily Dickinson Dissecting Brains" from *The Emily Dickinson Journal* (24, no. 2 [2015]: 46–70), which forms the foundation for Chapter 5, as well as my "Tapping the Wire: A Telegraphic Discourse," previously published in *American Literature* (88, no. 2 [2016]: 269–300) and adapted into sections of Chapter 6. Additionally, arguments and excerpts of readings from Chapters 1 and 6 have found their way into a German essay ("Medien der Begeisterung: Das Gehirn als technik- und literaturhistorisches Phänomen"), published in *Jahrbuch für Technikphilosophie* (2023: 17–34). All of these chapters have profited greatly from my respective anonymous reviewers as well as the assistance of journal editors and managing editors.

Finally, thanks are due to Hannah Doyle and Madison Zickgraf at Oxford University Press, as well as the editorial team at Newgen. This project has immensely benefited from two incredibly generous but anonymous reviewers. I am beyond grateful for their support of my work and their insightful comments have undoubtedly made *Writing the Brain* a better book.

From the bottom of my brain: thank you, all.

The First Century of the Brain

An Introduction

On January 23, 1866, the venerable physicians of the New Jersey Medical Society, the oldest organization of its kind in the United States, assembled in New Brunswick to celebrate their centennial. The occasion called for pathos. So before the society ended the first day of its meeting by "march[ing] in procession to Greer's Hall for dinner" (*Transactions of the Medical Society* 1866, 10), its president, Dr. Abraham Coles (1813–1891), stepped up to recite a poem. Skimming over the more than seventy pages of the published version, one can imagine the growling stomachs of the audience. But the topic of Coles's composition allowed for no brevity: He was reciting a hymn on the physiological grandeur of human anatomy. Titled "The Microcosm," the poem celebrates the human body as a model of earthly perfectibility; and, like "man" dominating creation, his own nature is dominated by a quite marvelous organ—the "big and billowy and powerful brain" (Coles 1866, 21):

> Here mounted, standing on the topmost towers,
> Up to the roof of this high dome of ours,
> With the Mind's Organ in our hands, what new
> Secrets of structure strike th' astonished view?
> A weird, and wonderful, and fragile mass
> Of white and gray—deserted now, alas!
> All knowledge quite razed out; no trace
> Of things which were: now mourns each happy place,
> Where frolicked once the Children of the Mind;—
> Of all the number, not one left behind. (Coles 1866, 38)

One has to wonder what the performance of these choppy verses sounded like. Did Coles, an accomplished church hymnist on the side, deliver them with the appropriate amount of pathos for the occasion? We might imagine him with

Writing the Brain. Stefan Schöberlein, Oxford University Press. © Oxford University Press 2023.
DOI: 10.1093/oso/9780197693681.003.0001

2 WRITING THE BRAIN

outstretched, cupped hands as if he were offering up an actual brain for his audience to inspect.

Pathos aside, there is something remarkable about this congregation of 275 renowned physicians attending a service celebrating the "packed and folded pulp of brain" (Coles 1866, 42). For a medical society older than the United States itself, and a profession that proudly harkened back to the days of heathen Galen, to consider this organ with an antiquity of millions of years "weird, and wonderful" and marvel at its "*new* / Secrets of structure," something rather groundbreaking must have happened. And indeed, it had: "The nineteenth century witnessed the establishment of neuroscience as a discipline" (Stiles, Finger, and Boller 2013, vii), and especially "early decades of the nineteenth century [saw] unprecedented collaborations between literature and science" on issues relating to the brain (Stiles 2019, 372). Within Coles's lifetime, the brain had morphed from a passive, inert tabula rasa temporarily inhabited by souls and animal spirits to a material system that *creates* mind—a "Mind's Organ" (Coles again) whose anatomical structure enables, limits, and defines the thinking self. What the brain offered up for the audience's examination was essentially a material act of introspection: In front of the doctor was a roomful of thinking brains marveling that a brain could possibly produce thought. The tools for this marvel were supplied by eager anatomists, and their findings can be observed in scientific and literary writing alike.

And indeed, much of the fascination of the nineteenth century for the brain is, if not flat-out determined by, then at least precipitated by the invention of actual tools and techniques. The craniologists Gall and Spurzheim had invented a way of hardening brain specimens in alcohol, enabling the doctors to extract individual fibers and centers, and thus leading them to conceptualize the brain as distinct systems and structures, not a uniform mold of simple matter (Wickens 2014, 138). As a result, the number of neuropathology autopsies increased, and dissection kits became mass-produced and progressively more standardized (McCampbell 2017). The newly invented stereoscope turned eighteenth-century sensationalism on its head by emphasizing the brain's role in visual apperception[1] before it became a vessel for mind-tourism and pornography (Cahan 1993, 178). And by the third decade of the nineteenth century (Singer 2011, 330), microscopes had become powerful enough to reveal how complex nervous matter truly was and how far removed it was from

---

[1] It did so by demonstrating clearly that the perceived image is generated in the brain, not in the eyes themselves—which complicated the seventeenth- and eighteenth-century conceptualization of sensory input as simple conveyance of truth: i.e., what we see is what the brain generates from diffuse input, not a rational reaction to clear images generated by objective senses.

notions of gray matter as passive clay, hollow hallways, or primordial jelly. In tandem, technological advances of the age replaced Enlightenment metaphors (tabulae rasae, corridors, strings, musical notes) in discussions about the brain with the language of atoms, batteries, and telegraph wires—ideas that emphasize a force *within*, not *without*.

Not only did the means to analyze the brain change, but so did the scope for such endeavors. From the 1800s onward, the number of specimens available to anatomically inclined physicians increased dramatically: With "lunacy" defined as a medical problem by the late eighteenth century, asylums for the insane sprang up all around Europe in the early 1800s (and twenty to thirty years later in the United States, often as a reaction against European standards of therapy and care) and supplied the basis for clinical psychiatry, comparative psychology, and (via postmortems) neuropsychological theories that tied behavioral anomalies to observable brain anomalies. In a similar vein, practices like medical grave robbing, dissection as a means of humiliating executed felons, and the system of slavery, which disallowed significant populations ownership of their bodies and brains (see Sappol 2002; Hutton 2015; Richardson 2009), enabled anatomy's rise to prominence at the dawn of the nineteenth century. Training followed suit: Within years, anatomy was a common subject not only in colleges (Rothstein 1987) but in secondary schools (Powers 1920, 46). Quickly, information on the brain became ample, and its many experts were optimistic about their ability to decipher the exact relationship between psyche and cerebrum.

And their results were, by any standard, marvelous. Had Dr. Coles's father ever looked for a drawing of the human brain, he may have found something like this:

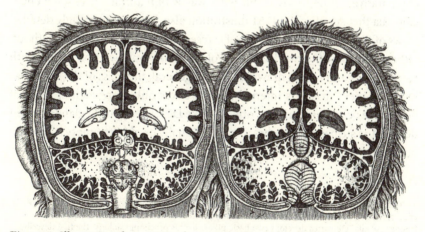

Figure 1 Illustration from 1751 of a cranial cross section (Diderot and d'Alembert 2017, 18:3:15).

4   WRITING THE BRAIN

Coles himself, on the other hand, would have encountered images that look much closer to how we know the organ today:

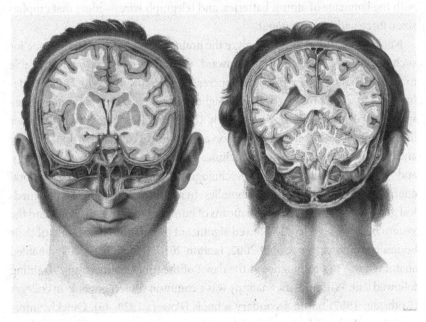

Figure 2   Illustration of a cranial cross section (Bourgery and Jacob 1844, 303).

The point here, of course, is not that one illustration was clearly performed by a more accomplished artist, but that eighteenth-century anatomy had comparatively little to say about what was *within* a brain. The same encyclopedia that supplies the 1751 illustration also features stunningly detailed views of the cerebrum, but all detail is reserved for depictions from without. Other intracranial artworks in the volume show hollowed-out skulls with certain veins and nerves drawn in, but, in general, the mass of the brain was just that: mere mass. Its internal structure was only important as far as it hosted the occasional minute "organ"—like Descartes's pineal gland[2]—often depicted as suspended in what may functionally be akin to empty space or a bucket of water.

---

[2] One of the conceptual underpinnings for choosing the gland was, of course, its singular existence in the cranium—a single locus for a unitary mind in an otherwise eerily doubled brain. Or as Hagner puts it: "Descartes develops two basic pillars for his psychophysiology: the first describes a

THE FIRST CENTURY OF THE BRAIN 5

The 1844 illustration, on the other hand, clearly shows concern about gray versus white matter, individual lobes and convolutions (and their slight hemispheric differences), and the exact topography of its fissures. If one looks at the results of the mapping of the brain by nineteenth-century anatomists, it becomes clear why the terra incognita of the 1751 drawing had to recede. In just the first half of the nineteenth century, we find the first descriptions/namings of the midbrain, substantia nigra, Broca's area, cavum vergae, fasciculus gracilis, respiratory center, claustrum, cingular gyrus, central sulcus, Schwann cells, and many others, as well as anatomical arguments for functional differences between lateral and medial geniculate as well as the dorsal and ventral roots of the spinal cord, and explanations for the role of the cerebellum in motor function. With naming and differentiating also came a drive to hypothesize *function*, resulting, in turn, in a functional particularization of the mind generated from such distinct structures, organs, and systems within the brain.

If one is inclined to hunt for epistemological breaks or ontological shifts, the mapping of the brain certainly qualifies. Whereas the 1751 encyclopedia (Figure 1) still had to endure public outrage and censorship over its uncouth depiction of human nature, by the time of Coles's centennial versifications, lumpy brains had become a topic for polite conversation and the occasional celebratory poem. In the early nineteenth century, as Robert Young rightly observes, the understanding of the human psyche moved away from philosophical "epistemology" and began "its close relations with neurophysiology" (Young 1970, 190). But epistemology still has a way of creeping in, and even this "new" brain was far from safe. As Alfred Whitehead (from a 1920s perspective) notes:

> During the seventeenth century there evolved the scheme of scientific ideas which has dominated ever since. It involves a fundamental duality, with *material* on the one hand, and on the other hand *mind*. In between there lie the concepts of life, organism, function, instantaneous reality, interaction,

necessary connection between immateriality and indivisibility of the soul; the second one—and this is where he extends his notion of unity from immateriality to matter—a singular interaction space has to be located that only exists in one place. This means that only those structures can even qualify that cannot be found in both halves of the brain (and thus exist twice). . . . The most important consequence of Descartes's theses is that the brain as a whole cannot be conceptualized as the seat of the soul because it is constituted by two, roughly identical halves" (2000, 27, translation mine). See Chapter 2 for an exploration of this duality.

order of nature, which collectively form the Achilles heel of the whole system. (1953, 71)

The anatomical brain science of the early 1800s—after the mind had abandoned abstract philosophy for biology and before it would rekindle its love for dualism via modern psychology (then as primarily a methodological dualism)—was *all* Achilles heel, a potentially fatal weak spot to all notions of metaphysical minds. Stripped down to its "pulp," the "new" biological mind presented its owners with a number of quandaries about free will, the (im)mortality of the self, the dominance of mind over body, self-control, and rationality. There was much to ponder.

While "the localization of mental functions was not a generally accepted fact for a long time after its inception," Michael Hagner notes, "it had such an impact that everyone had to respond to it" (2000, 10, translation mine). Consequently, debates about brains were by no means confined to academic journals or physicians' meetings. The 1800s, as Sonja Boos reminds us, was a time "when the boundaries between the sciences and the humanities and their respective representational modes were far more fluid than they are today" (2021, 213). A glance at the totality of texts of the century currently digitized by the corporate hive mind of Google clearly shows the scope of the century's brain fever:

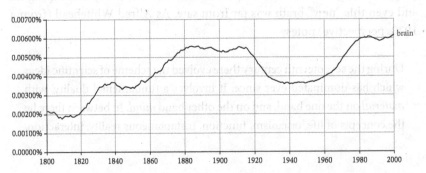

Figure 3 Relative usage of the word "brain" in the English corpus of Google Books for the years 1800 to 2000 (smoothing value of 5, Google Ngram Viewer).

THE FIRST CENTURY OF THE BRAIN   7

Between 1800 and 1880 cerebral talk almost doubled in print (before stagnating, and then dropping during the height of psychoanalysis).[3] The brain was trending, and it did so in England and the United States alike.

* * *

Both the science and literature of the day fluidly moved back and forth across the Atlantic, with British mind-materialists fleeing to the United States, American heroic alienism being debated in the United Kingdom, and neurophysiological research as well as spectacular anecdotes about psychiatric patients circulating with little to no transmission lag in the periodical spheres of both nations. If you wanted to read about a man surviving cerebral impalement or a woman's sleepwalking brain compelling her to drown herself, you could do so with equal ease in Sheffield or Philadelphia. Essentially nonexistent transatlantic copyright restrictions and a culture of reprinting only intensified scientific and popular discourse about the brain on both sides of the pond.[4] Of course, English-language works rather quickly spread through other (Western) nations as well, but the selective filter of translation and the more distinct nature of national literatures of different languages account for a somewhat narrowly transatlantic, anglophone focus of the following chapters—albeit with the full acknowledgment of this being a somewhat heuristic distinction, not a clear-cut, categorical difference. Non-English-speaking scientists make their appearance where they left their traces on British and American texts, but the deep entanglement of these figures with their native cultures will not figure prominently in these pages. To discover what Samuel von Sömmerring's Germany thought about the brain (see Ione 2016, 141–49; Siegert 2003, 282–94; Boos 2021), how Philippe Pinel's France debated diseased minds (see Thiher 1999, 131–54), or how Spyridon Mavrogenis's Ottoman Empire engaged with mind-materialism (see Anogianakis 2014) would certainly speak to the themes and thinkers of this book, but such questions will have to be addressed elsewhere.

At least in the English-speaking centers of print culture, there was no escaping the brain: it was in newspapers and monthlies; it was in schools and lyceums; it was debated on podiums in front of crowded amphitheaters; it was

---

[3] Even when confined to Google's "Fiction" corpus of the time, the development is essentially the same.

[4] Perhaps as a nod to British cultural hegemony in the early 1800s, most of the anglophone scholarship on the material brain in these years (such as the essential research produced by Victorianists Anne Stiles, Alan Richardson, Shelley Trower, and Gregory Tate) has focused on England.

8 WRITING THE BRAIN

performed onstage; it was just around the corner in the form of an asylum, a phrenological parlor, or a medical college; it was in the family as mental illness or as the object of study for a relative in the medical profession—and it certainly found a home in the literary works of the day, whether it was discussed outright or roamed the margins.

It should come as no surprise, then, that Emily Dickinson learned from anatomical drawings of opened skulls as a teenager and returned to these images in her poems, or that Samuel Taylor Coleridge was well versed in the proto-neuroscience of his day and debated its theories throughout his adult life, or that Walt Whitman glued charts of the brain into his daybooks and urged himself in handwritten notes to "*Take Notice*" of "*The Physiology of the Brian from a Literary Life*" (c. 1859), or that Emily Brontë had much to say about the dual structure of the human cranium and speculate about its effects on the psyche.

Still, while these *shouldn't* come as a surprise, cranial contexts like these nonetheless often do. While psychological criticism has a long history in literary scholarship, this peculiar neurophysiological line of thinking of the 1800s remains underappreciated. Seldom has scholarship treated these theories as more than a curious foreshadowing of psychoanalysis, mainstream psychology, or modern cognitive narratology. Their often quite radical implications are rarely allowed to stand on their own. Some theories, like phrenology, are so carnivalesque from today's perspective that when they have attracted significant scholarly attention (for instance, with regard to Whitman's thinking on queer identity), their treatment often seems to require, in hindsight, a scornful dismissal of the science and/or an emphasis on writers having "moved beyond" or "fundamentally reinterpreted" it. Other concepts, like "dual brain" theories, have remained marginalized in the history of science more broadly and thus tend to elude many literary scholars as well.[5]

This relative lack of scholarship is not due to the nature of the subject at hand. Questions of mind and psyche have a long and intricate relationship to literary studies. The modern period of psychology, for instance, that began roughly around the 1880s has received ample scholarly commentary. This makes sense, considering how its theories more readily lend themselves to Foucauldian readings of (self-)discipline or psychoanalytical

---

[5] Exceptions are the crucial scholarly contributions by historians of science C. U. M. Smith, Anne Harrington, and Harry Whitaker.

THE FIRST CENTURY OF THE BRAIN   9

analyses of displacement and repression. In more narrowly neurophysiological terms, this period has been variously framed as the "age of nervousness" (Pietikäinen 2007), a "golden age of memory" (Nalbantian 2004), the "golden age of hysteria" (Bondevik 2010, 183), or the "age of insight" (Kandel 2012), generating a wealth of interpretations of the formation of the modern subject at the nexus of psychological introspection and narrative self-fashioning. Similarly, media theory—a crucial tool to comprehend the material brain as simultaneously a medium and an actant[6]—has also found itself enamored with mostly the latter decades of the century and in turn morphed the whole period into the age of Hermann von Helmholtz (1821–1894) and his technophile theories on nervous transmission and mediated subjectivity. Preceding decades, however, have yet to see their own, comparable scholarly renaissance. Especially excellent scholarly volumes on particular writers or concepts exist—particularly the work of Anne Stiles and Sally Shuttleworth should be emphasized here—but the broader scientific-literary culture of the period remains understudied and unnamed.

Focusing on these earlier decades, of course, does not diminish contributions to the science (or the academic study thereof) of later periods, but helps us paint a richer picture of their historical, scientific, and literary foundations. In recovering the brain infatuation of the early 1800s, *Writing the Brain* aims to expand our appreciation for the cultural-scientific advances, errors, and quandaries of its day. It does not sketch out a counterhistory but is additive: It hopes to be an opening salvo for a border consideration of the sciences of the first decades of the nineteenth century by literary scholars— which we then may call the "first century of the brain."

My use of terms like "early neuroscience" or "proto-neuroscience" in this context, it should be noted here, aims both to underscore existing and well-documented lineages in the history of science and to find a usable umbrella term for a heterogeneous grouping of anatomists, phrenologists, practicing physicians, psychometrists, niche philosophers, crude materialists, science writers, and pop psychologists. Interrogating the brain in the pre-professional era of medicine, this field naturally looks quite different from the interdisciplinary field of modern neuroscience that came to exist in the mid-twentieth century. As Anne Stiles reminds us, "mental science disciplines and the modern professional networks that supported them existed only in embryo"

---

[6] I.e., as an "agency or means of doing something"—an "intervening substance" that is subject and object *simultaneously* (*Oxford Living Dictionary*, s.v. "medium").

10   WRITING THE BRAIN

at the time and often "incorporated discoveries from fields we would now consider pseudoscientific" (2021, 40–41). This does not, however, mean that their now-obvious incorrectness renders such discoveries mere dead ends or theoretical detours. This holds true even in cases as seemingly absurd as phrenology, which constituted the first modern theory of brain localization. Ultimately, in the same way as we might call Benjamin Rush an "early epidemiologist" without denigrating the modern field of study and its later-coined name (and without defending Rush's bloodletting), I believe we may justly call Arthur Wigan an "early neuroscientist." This book, then, uses terms like "brain science" or variations of "proto-neuroscience" when attempting to dissect the ambition, intent, and scope of these early thinkers, while maintaining a focus on foundational insights into the functional anatomy of the brain made during the decades in question, such as hemispheric lateralization, brain localization, or neural theories of cognition.

The "first century of the brain," as these pages present it, is, then, roughly framed by the years 1800 and 1880. Of course, as differentiating "ages" is always a problematic endeavor—with literary production so full of "anticipating" and "echoing" texts that rarely fall in line neatly—the chronological delineators for the following chapters ought to be seen as guideposts, not barbed wire fences. The first date marks the height of Benjamin Rush's lecturing activities on the mind, the publication of Pinel's groundbreaking *Traité médico-philosophique sur l'aliénation mentale ou la manie*, and the fateful first meetings of Johann Spurzheim and Franz Joseph Gall. It marks the dusk of the "golden age of introspection" (Lyons 1986) and the dawn of comprehending the mind via the brain, a process requiring the aid of physicians, not (just) philosophers. The latter bracket to these chapters, the year 1880, on the other hand, sees empirical psychology wrestle *Deutungshoheit* (interpretive hegemony) away from its more anatomically inclined colleagues. One of the foundational figures in this move, Wilhelm Wundt (1832–1920), began in 1880 to publish his multivolume work on logic and laid the foundation for psychological introspection as an empirical method. His famous psychological laboratory, too, had just opened its doors in Leipzig (1879). Within a few years, the first doctorates in the new field of psychology were awarded and professorships established—including for William James at Harvard, who had previously lectured merely on "anatomy and physiology." At the same time, the "drifting apart [of anatomy and physiology] that began after the 1840s" is embodied by figures like Helmholtz and Emil du Bois-Reymond (Hagner 2000, 72). Between 1800 and 1880, it was

THE FIRST CENTURY OF THE BRAIN   11

the organ of the brain—its structure, form, and constituent matter—that became the beacon of scientific and literary inquiry that this book will focus on.

* * *

The following chapters trace the first decades of the material mind and its literary effects through various thematic discourses. These discourses, each mapped to a different literary effect, all describe specific developments stretching from the early to the late 1800s and thus at times parallel, overlap, and feed on each other. What unites them is a shared trajectory from dualism to monism, soul to matter, cognitive unity to complexity, and subjecthood to cranial subroutines. *Writing the Brain* begins with an exploration of mind-materialism in the *Naturphilosophie*-style ideas of the British Romantics, then moves into early theories of cerebral localization proposed by dual-brain theory and phrenology, contextualizes the racialization of the brain in psychometry, visits depictions of insanity as metacognitive acts, and concludes by narrowing in on the shared metaphorical nexus between early neuroscience and Victorian communication media. The goal is to venture beyond influence studies of particular authors and instead show how science and literature in tandem worked through questions posed by the "new" brain. All the authors I discuss repeat the awe-filled gaze of Dr. Coles staring at the brain, interrogating its meaning: What are the limits of free will, or can there even be free will? If the self is spread over distinct parts of the cranium, what happens to the mind (and the self) when the brain is damaged or destroyed? Can a material brain, adhering to natural laws, really be rational? Is the mind fully in control of the brain, or are there parts of the brain outside of its reach?

The first chapter, "Nature's Mind and Mind's Nature: Romantic Cognition Between Harp and Atom," traces the shifting understanding of the brain by the British Romantics through an examination of the metaphor of the Aeolian harp, the prime image used to describe a brain reacting to nature. Centering on readings of Coleridge, Wordsworth, and Shelley, this chapter discusses the turn-of-the-nineteenth-century debate on the shortcomings of philosophical theories of mind (especially the Associationism of David Hartley) and observes how literary and scientific writers move from eighteenth-century theories of the brain as vibrating cords to a chemically inflected concept of mind as an atomic property of biological matter. This shift in reading the brain, this chapter argues, leads to a double-edged appreciation of the self by these poets: It elevates the mind to a structuring force in

12 WRITING THE BRAIN

nature, while at the same time challenging the boundaries that uphold the thinking, autonomous self.

The subsequent chapter, "Split Brains, Doubled Minds: The Gothic's Bicameral Vision," moves from these more basic, philosophic-scientific issues of mind-materialism and turns toward first attempts to describe the functional specialization of the brain. Since perhaps the most striking anatomical feature of the human cranium is its bicameral structure—its two halves that make it possible in the English language to speak of a person's "brains" in the plural—this chapter traces the effect of "dual mind" concepts in the second wave of the Gothic in literature. Because the dual hemispheres of the brain provide an intuitively persuasive anatomical argument for subconsciousness, its advocates (most notably Henry Holland and Arthur Ladbroke Wigan) ascribed opposing mental states to each hemisphere, with Wigan going so far as proposing a completely separate but suppressed mind to inhabit an anatomically inferior side of the brain. While the story of Dr. Jekyll and Mr. Hyde is perhaps the most pronounced (and belated) example of this idea in English literature, many of the early Gothic's oddly doubled characters also come to mind. This chapter thus begins by tracing the prehistory of Wigan's ideas in the popular fascination with sleepwalking and its cranial underpinnings through the work of Benjamin Rush, Charles Brockden Brown, Alfred Tennyson, and Edgar Allan Poe, before turning to echoes of Wigan's theories in the mirrored protagonists of Emily Brontë's *Wuthering Heights* or the cranial doppelgängers in Herman Melville's "Benito Cereno." The brain, this bicamerally inflected Gothic suggests, is not a coherent whole but a split creature: What is experienced as the "self" inhabits only parts of it, with the other part potentially subverting the "self's" every decision.

The third chapter of this book, "Skulls and Society: Reading the Mind as a Multi-Organ Entity," addresses perhaps the best-known early theory of functional brain specialization: phrenology, the theory that the brain is a multi-organ structure and behavioral traits can be mapped onto different regions of it (and measured through the skull). This chapter begins with the story of Phineas Gage's infamous accident in which he lost parts of his brain, yet lived through it in near-perfect health, but with pronounced changes in personality and behavior. The case became a famous cause célèbre for early proponents of brain localization. Many literary (proto-)Realists and ardent believers in the phrenology of the day (specifically Charles Dickens and Walt Whitman) contemporaneously began using the trope of behavior change as a result of "blows upon the head" to an almost excessive degree. Given the determinist gist of phrenology—it quite openly denied free will—abrupt psychological changes without the application of external force were now simply no longer a "realistic" option for novelists. Following this observation, the chapter branches outward to discuss the societal vision of phrenology as developed by George Combe, who turned a rather obscure science into perhaps the first modern self-help craze. Combe's phrenology preached a vision of society as fundamentally self-regulatory and ultimately static; it was a vision adopted by mid-Victorian Realists and turned the reformist impetus of their works into a call for restoring balance to society instead of overcoming its structuring injustices. The American Fowler brothers—friends and supporters of Whitman at the time he was writing *Leaves of Grass*—adapted Combe's theory and made it their own, and the chapter concludes by reading how their cranial theory of biological republicanism and American self-esteem permeates the work of Whitman, one of their most ardent literary customers.

The fourth chapter, titled "Cranial Reconstruction: Racialized Brains and the Psychometric Real," accounts for the dominance of early psychometrics in the form of racial comparative craniology, in particular the measurement of brain volume. It begins with the reading of Turner's *Confessions* to outline Black genius as a challenge to ubiquitous theories of racial cognitive difference, leading into an examination of Great Man theory in advocacy works like those of William Wells Brown in relation to the reformist impulse of the otherwise racially conflicted and complicit field of phrenology. These moments, I argue here, outline the emergence of a trend toward psychometric thinking that is exemplified by the medical testing on Black brains by the Union Army. In fiction, it suggests a raced "psychic coherence" (Bersani 1976, 6) on the level of characters and the reader. It is an act of cultural midwifery of sorts, creating a psychometric gaze that then infects the birth of the Realist novel, as exemplified through a brief media-theoretical explication of *Miss Ravenel's Conversion* as "stereographic fiction."

The penultimate chapter, "Rattle-Brained: Insanity as Material Metacognition," adds to these observations by analyzing literary renditions of insanity as a means for authors to debate the material nature of mental life. This chapter describes an anatomically justified psychology that draws from and simultaneously critiques medical notions of insanity and the practice of institutionalization. Focusing on the work of Robert Montgomery Bird, Wilkie Collins, and Emily Dickinson, it proposes that these authors narrate

14   WRITING THE BRAIN

insanity as a metacognitive act (an instance where a speaker can observe him- or herself think) in order to interrogate the psycho-physiological limits of thought as such. Drawing from the work of Benjamin Rush, founding father both of the United States and of American psychiatry, and Pliny Earle (friend of the Dickinson family and Rush detractor), the chapter argues that these writers laid the conceptual groundwork for the mimetically unreliable narrators of modernity by following the brain science of their day to its logical extremes and discovering there the limits to their own biological mind and, consequently, to their authorial voice.

Writing the Brain concludes with the chapter "The Telegraphed Brain: Wires as Proto-Neurons," which traces the long history of the neuron doctrine via the cultural phenomenon of the telegraph in British and American literature. At a time when many scientists still considered the majority of the brain to consist of a simple "primordial mass," the telegraph—as metaphor and communication medium—proposed a different take on the material nature of thought. Even decades before neurons could be observed in the brain, the now-common experience of transmitting "thought" over wires teased the cultural imaginary of the mid-1800s to conceptualize the brain as a system of connected wires—proto-neurons of sorts that stored and forwarded information along branching but predetermined paths. From sensational periodical fiction to the accounts of a prison camp survivor, a multitude of writers, engineers, and physicians (both trained and amateur) in these decades began to translate the Romantic notion of brain-as-harp into the technophile metaphor of the telegraph-brain—a metaphor that, in turn, laid the groundwork for one of the most important insights about the material structure of the brain: its electrochemical, neuronal nature.

Reading "literary texts alongside scientific theories," N. Katherine Hayles notes, serves to "entangle abstract form and material particularity such that the reader will find it difficult to maintain the perception that they are separate and discrete entities" (2010, 22–23). As such, this book builds on Hayles's notion of literary works as "local habitation" for philosophical and scientific discourse in what is, ultimately, a problem of embodiment. Embodiment here, of course, serves as a fascinating prehistory to Hayles's disappearing bodies of cybernetics; it follows a trajectory of autarkic subjects becoming bodily brains to the logical extreme that then lays the monistic groundwork for the resurgent dualism that Hayles traces. The brief moment in time, when the mind-body dualism of enlightened humanism was epistemologically overcome and the data-body dualism of burgeoning posthumanism was still decades away, stands at the center of this book.

As these preliminary notes indicate, media studies, in particular in the vein of so-called German media theory, will play an outsized role that might seem curious at first. Media, of course, in this context, does not simply mean "technology" but describes "symbolic operators" and "drill practices [that] were located at the base of intellectual and cultural shifts" (Siegert 2015, 2). In this book, such framing becomes a helpful tool to unravel the complexities encoded into the cultural, material object of the brain, which is all too often read as mere nature or pure spirit. As Friedrich Kittler put it, the 1800s saw a disintegration of the brain "into a large number of subroutines, which physiologists could localize in different centers of the brain and engineers would reconstruct in multiple machines" (1999, 188).[7] Kittler's mistake here, of course, is that he is nonetheless occasionally treating medical insights as culturally unfiltered expressions of the Real—a "truth" unearthed by experts. As the following pages will demonstrate, this is shortsighted. Instead, the brain itself became a medium, tied into a discursive structure that sees culture and science intermingle, co-creating both medical "truth" and its resulting apparats—textual, technological, and otherwise.

This structure Kittler deemed an *Aufschreibesystem*[8]—a network of interwoven cultural techniques, attitudes, and norms that focuses on a particular cultural icon (technological or otherwise) to make visible the cultural system of meaning-making that constitutes it and that might otherwise remain obscured. Tracing this discourse network of the brain is the task of the following chapters. Such a form of reading thus has to be interested in

---

[7] Siegert in his monumental *Passage des Digitalen*, for instance, has much to say about various techno-medical innovations and their relationship to cognition—all with the ultimate goal, however, of demonstrating that cognition itself is, at best, material resistance of electronic relays, conceptualizing itself into insignificance in a larger process of a "convergence of inscription and matter" (2003, 13). To Siegert the human mind disappears alongside the difference between scribe and switch in the digital-electronic realm of pure, deterritorialized writing. One of Siegert's prime enemies in *Passage* is the idea of the "transcendental unity of apperceptions" (2003, 288) of the eighteenth-century *sensorium commune*—which, he argues, Kant destroyed as pure discourse (2003, 295): "In the ultimate dispersal of cognition, this failing synthesis, reappears the excluded body of inscription" (2003, 292). Siegert's description of the brain as "that which is behind the eyes" is telling (2003, 368). In dismissing brain science before it became a science proper, and only revisiting the brain in a chapter on the 1900s (2003, 369–400), Siegert is bracketing off the epistemological event that Benn observes in the birth of modern neuroscience in the 1800s. Instead, Siegert attempts to tie Benn to Helmholtz (2003, 368), ignoring Benn's own celebration of the "first half" of the 1800s that introduced this book.

[8] Often translated as "discourse network," but literally "recording system" or "transcription system."

16 WRITING THE BRAIN

curious asides, "arcane sources" (Siegert 2015, 3), and freak science as much as it finds value in novel pathways into highly canonized works by major writers of science and other fictions.

Besides its conceptual implications, this orientation affects the method and style of this book. If one is to trace the mutually constitutive, moving parts of the *Aufschreibesystem* brain, a mere set of close readings, chronologically ordered, will not suffice. Hence, discursive reading strategies are marshaled to understand the cultural genesis of the modern brain as a set of cultural developments that spans genres, writers, and continents. This book thus traces "subroutines" within the brain through their various cultural moments. Participation in the discourse around the brain, as laid out here, is never unidirectional in the sense that an "influence study" might imply but discursive—less interested in "origins" and "reception" than in mutually constitutive metaphors, slippages, and obsessions. Some of these discourses lean toward the American context (Chapter 5), some toward the British (Chapter 1), but all relish multidirectional *exchange*. This study's interdisciplinarity and transatlanticism are pragmatic—it follows networks of writing, reading, printing, and reprinting wherever they appear.

* * *

The specific networks that birthed the first century of the brain largely receded from popular reading with the birth of modern psychology. The cranial optimism that lingered behind even the darkest of literary treatments of the "new brain"—the idea that the human mind will soon be perfectly legible via the anatomy of the brain—came to a rather abrupt halt. The tools of the age, while able to usher in such radically novel insights into the organ, were still much too blunt to really make use of its puzzling complexity. Even in 1866, Dr. Coles seemed to anticipate the ultimate frustration of his "pulpy" ambitions:

> Where [these children of the brain][9] had their high dwelling, we, in vain,
> Seek in this packed and folded pulp of brain:
> Judged, by the ignorant regards of sense,
> How mean! by heights of function, how immense!
> ...

---

[9] In the stanzas leading up to this one, Coles specifically identifies these as Reverence, Memory, Imagination, and Conscience.

THE FIRST CENTURY OF THE BRAIN 17

So high so hidden—from whose secret tops
Keener than needles, trickled the first drops
Of rising rivers, flowing silently
Into the cerebral deep drainless sea,
From which, as from a mighty fountain-head,
Life's crystal waters everywhere were spread,
Coursing in liquid lapse through Channels White,
Swift as the lightning, stainless as the light,
Conveying to each atom of the whole
Volitions, animations, power and soul. (1866, 42–44)

After more than half a century of cerebral optimism, the true meaning of the brain was still "so hidden," its "secret tops" eluding the tools of the scientist, rendering his/her ambitions "in vain."

If we think back to our image of Coles's outstretched arm, cradling an invisible brain in front of his fellow physicians, we might conclude they may have dropped the ball—or, to speak with Emily Dickinson, "dropped [the] brain."[10] Actual brain-dropping, of course, did take place. A doctor/phrenologist famously tasked with comprehending Whitman's mind by performing a postmortem on his brain accidentally did just that, leading to outrage among the poet's disciples (Burrell 2003; Gosline 2014).[11] Before that, the doctor had found very little in the folds of the poet's brain that he considered remarkable—an indictment not of the poet, it seems, but of the science employed.

Reading the mind in the brain seemed to grind to a frustrating halt and splinter into more specialized disciplines, often outside of the purview of popular culture. In the literatures of the following decades, the psyche would outcompete the brain for prominence. In 1873 a key voice in early psychology, Alexander Bain (1818–1903), would propose "guarded materialism" to replace the "crude" materialism that had reigned almost unchecked in the previous decades of the century (1873, 140). Bain, too, grounded his theories in a "*double-faced unity*" of "the mental and the physical": "one substance, with two sets of properties" (193). But instead of using the latter to understand the former, he suggested a reversal that would become the cornerstone

---

[10] "I've dropped my Brain — My Soul is numb —."
[11] As poet Thomas Lux imagines it: "At his request, after death, his brain removed / for science, phrenology, to study, and / as the mortuary assistant carried it / (I suppose in a jar but I hope cupped / in his hands) across the lab's stone floor he dropped it" (1997, 112).

18 WRITING THE BRAIN

of modern psychology. In place of reading the brain to get to the mind, he argues that the mind as such can be empirically assessed on its own, with the assumption that the results thus gleaned surely correspond on a physiological level as well. Still, comprehending the exact nature of this correspondence was no longer required. Within half a decade, Wundt had perfected his theories on scientific introspection, and brain anatomy was once again logically and methodologically removable from analyses of the mind. The Achilles heel of material cognition that Whitehead was later to observe was by the late 1870s being quickly bandaged up by the new, pragmatic dualism of modern psychology.

Still, from today's perspective, when phrenology and related pseudosciences are experiencing an uncanny resurgence in business consulting (under the terms "physiognomic consulting" or "face profiling"), dual process theories in modern cognitive neuroscience clearly divulge echoes of nineteenth-century "dual brain" concepts,[12] and the material nature of psychiatric illness is again hotly debated (cf. Karson 2014), it may be time to pick up the pieces of Coles's brain. New waves of neuroscientific research—and their at times problematic ties to psychologically manipulative marketing, political strategizing,[13] and the pharmaceutical business—are carrying the material brain again to the forefront of public discourse (and by now surpassing even the "brain fever" of the nineteenth century, if we are to believe Google's corpus). A glance back to the birth of the material mind may then act both as history lesson and as corrective.

---

[12] Including some "bold" scientific voices that embrace the idea of two distinct minds in one brain (St. Evans and Frankish 2009, v).

[13] Cambridge Analytica, the company entangled in allegations of Russian interference in the American presidential election of 2016, notoriously used a human brain as its company logo.

1

Nature's Mind and Mind's Nature

Romantic Cognition Between Harp and Atom

I feel the spell of long-forgotten yearning
for that serene and solemn spirit realm,
and like an Aeolian harp my murmuring song
lets its uncertain tones float through the air.

— Goethe, *Faust*

These syllables that Nature spoke,
And the thoughts that in him woke,
Can adequately utter none
Save to his ear the wind-harp lone.

—Ralph Waldo Emerson, "The Harp"

Harp Strings

After having composed the most successful piece of poetry of the English Romantic period (Cochran 2014, 1; White, Goodridge, and Keegan 2006, 27)—a feat allegedly performed "*in his head*" long before taking up the pen to record it (Cromwell 1859, 129)—Robert Bloomfield was looking for another endeavor to sustain himself and his ailing family.[1] With his subsequent poetry failing to live up to the sales of his breakout hit *The Farmer's Boy* (1800), the rural Suffolk poet and women's shoemaker moved to London and, in 1806, took up a new kind of poetic business. Besides offering up to the masses his Romantic mind and soul carved onto paper, he would now build machines to do the necessary labor for him: Aeolian harps—the prime

---

[1] The Bloomfield household was affected by a number of "physical and mental health" issues (White, Goodridge, and Keegan 2006, 68). By 1814, Bloomfield had lost his wife and daughter and was suffering from what sources call either a "poor mind" or outright "insanity" (Sampson 1979, 536).

Writing the Brain. Stefan Schöberlein, Oxford University Press. © Oxford University Press 2023.
DOI: 10.1093/oso/9780197693681.003.0002

20 WRITING THE BRAIN

technological proxy for the poet of his day, translating the natural sublime into form and art. And by all accounts, Bloomfield's business with these poet-automata was initially booming (White, Goodridge, and Keegan 2006, 22).

Still, Bloomfield's investment in this "central metaphor for the Romantic imagination" (Byerly 1997, 45; see also Dix 1988, 288; Erhardt-Siebold 1931–32, 357–63) was more than a clever marketing strategy by an early trend adopter. Bloomfield not only introduced some essential mechanical changes to the device that significantly improved its sound quality (Goodridge 2002) but also theorized and anthologized it as the instrument par excellence for Romanticism. His short 1808 volume *Nature's Music* (reprinted in 1824) performs a task that goes beyond what its humble subtitle (*Practical Observations, and Poetical Testimonies*) suggests and instead sketches a collage history of the musical box that positions it at the very beginning of a line of thought that that would find its logical conclusion in the "sonorous bodies" of Romanticism (Bloomfield 1824, 110). And like any decent mythos of Western poetics, Bloomfield looked toward Greece for its origin story.

Whereas philosopher of media Friedrich Kittler would find himself two hundred years later, with "tears in his eyes," yelling at the rocky cliffs of the Sirenusas in his quest for the origin of poetry (2009, 57), Bloomfield locates it a mere 120 nautical miles south of the peculiar German's chosen site. To the English poet-inventor, it was not the Sirens playing the cithara that ushered Western thought into modern harmony but another acquaintance of Odysseus: the "poet-made god" of the Aeolian Islands and the harp named after him (Bloomfield 1824, 97–98). Quoting from John Hawkins's *General History* of music, Bloomfield's pamphlet nonetheless mentions the harp's Siren-like qualities, comparing its notes to a "chorus of voices at a distance" (1824, 108)—a heavenly choir freed by the ability of the harp to tease out the "seven degrees of sound" (1824, 112) inherent to air itself via its qualities as an "air prism" (1824, 114). Much in the same way as an optical prism discloses the component colors of light, the pamphlet argues, the Aeolian harp creates the purest harmony possible by separating the primary notes embedded in the wind itself. The Aeolian mode then constitutes a "*harmonics of the harmonics*" (1824, 107)—an originary artistic system that structures all lesser arts that sprang forth from it.

Bloomfield hence sides with Charles Bucke's claim that "nothing can be deemed natural but what proceeds from the actual principles of nature" and concludes: "We may safely pronounce the Eolian lyre to be the only natural

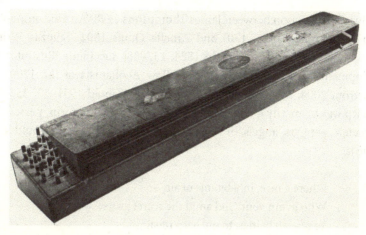

Figure 4 Wind-harp made by Bloomfield.

instrument of emitting harmony" (Bloomfield 1824, 121). In *Nature's Music*, the Aeolian harp is not a recently rediscovered gimmick originally designed by Athanasius Kircher in the 1600s but a thing of mythical antiquity. From the Homeric gods of Odysseus, the harps of Memnon (Bloomfield 1824, 124; see also Dix 1988), and the divine breath that "brought music out of the harp of David" in the Talmud (Bloomfield 1824, 103) to the present day, the instrument whispered in the language of Nature writ large (cf. Minssen 1997). Wind harps emitted, "without a metaphor, the music of inspiration" (Bloomfield 1824, 108), which so "wonderfully dispose[s] the mind for the most romantic situations" (Bloomfield 1824, 119).

Even the best of poets—and many of his contemporaries (from Coleridge to Wordsworth) would have counted Bloomfield among them (Cochran 2014, 3)—could thus not but fail to live up to this lyre's primordial poetics. Still, the wind harp promised too much to the Romantic imagination to not make imitation tempting, and soon almost every major writer of the time— be it Byron, Shelley, Wordsworth, Thoreau, or Emerson—composed poems debating the device. Reading through Bloomfield's early anthology (which is ordered chronologically), one can notice a fascinating shift around the turn of the nineteenth century in what the Aeolian harp communicated to its poets: a shift from nature transmitted to the nature of transmission that laid the groundwork triumphal procession of this "Radio of the Romantics" (Erhardt-Siebold 1931–32, 357) through the anglophone literature of the period.

22 WRITING THE BRAIN

A brief comparison between James Thompson's 1748 "An Ode on Aeolus's Harp" (Bloomfield 1824, 130) and Amelia Opie's 1802 "Stanzas Written Under Æolus's Harp" (Bloomfield 1824, 137–38) succinctly illustrates this development. To Thompson and his fellow Aeolianists of the 1700s (cf. Brunström 2004, 84–85; Matlak 2002; Erhardt-Siebold 1931–32, 358–60), the harp was primarily a messenger: It channeled air into form, personified as specters, zephyrs, angels, or a maiden's ghost, so the speaker could interrogate it.

> Ethereal race, inhabitants of air,
> Who hymn your God amid the secret grove,
> Ye unseen beings, to my harp repair,
> And raise majestic strains, or melt in love.
>
> Those tender notes, how kindly they upbraid!
> With what soft woe they thrill the lover's heart!
> Sure from the hand of some unhappy maid
> Who died of love these sweet complainings part.
>
> But hark! that strain was of a graver tone,
> On the deep strings his hand some hermit throws;
> Or he, the sacred Bard, who sat alone
> In the drear waste and wept his people's woes.
>
> Such was the song which Zion's children sung
> When by Euphrates' stream they made their plaint;
> And to such sadly solemn notes are strung
> Angelic harps to soothe a dying saint.
>
> Methinks I hear the full celestial choir
> Through Heaven's high dome their awful anthem raise;
> Now chanting clear, and now they all conspire
> To swell the lofty hymn from praise to praise.
>
> Let me, ye wandering spirits of the wind,
> Who, as wild fancy prompts you, touch the string,
> Smit with your theme, be in your chorus joined,

NATURE'S MIND AND MIND'S NATURE   23

For till you cease my muse forgets to sing. (Thompson, in Bloomfield 1824, 130)

In Thompson's poem, the wind harp is a translational interface, allowing the poet glimpses into a world of ethereal beings. His act of listening in turns into a spiritual visitation,[2] a lesson in the deep history of melancholia that connects a weeping Jeremiah ("the sacred Bard, who sat alone") to the "unhappy maid / Who died of love." What unites these figures is a soulful yearning to (re)join the angelic choir by which "such sadly solemn notes" become the "awful anthem" of a heavenly afterlife.

If we then look at Opie's poetic "talking back" to Thompson—given the numerous echoes of his "Ode," it was clearly on her mind when composing these "Stanzas"—we find her complicating his angelic reading of the harp. After calling all melancholy souls to the instrument so that "Each woe its own appropriate plaint may hear," she continues:

> But hark! in regular progression move
> Yon silver sounds,[3] and mingle as they fall;—
> Do they not wake thy trembling nerves, O Love,
> And into warmer life thy feelings call?
> ...
>
> But the wild cadence of these trembling strings
> The enchantress Fancy with most rapture hears;
> ...
>
> O breathing instrument! be ever near
> While to the pensive muse my vows I pay;
> Thy softest call the inmost soul can hear,
> Thy faintest breath can Fancy's pinions play.

[2] Emily Dickinson, who had been given an Aeolian harp by her cousin (Dickinson 1971, 179), would later parody this notion in her 1860s poem "The wind tapped like a tired man" (Dickinson 1999, 621), which turns the personified breath of Aeolus into an awkward visitor, unable to even properly sit on a couch (due to his inconvenient lack of a physical body), whose "Speech was like the Push / Of numerous Humming Birds at once / From a superior Bush -." See also Loeffelholz 1991, 129–30.

[3] "Strings covered with silver wire I have tried in various ways, but am not prepared to say whether they perform their part. Perhaps the metallic covering is not adapted to the action of the breeze" (Bloomfield 1824, 118).

24 WRITING THE BRAIN

> And when art's labour'd strains my feelings tire,
> To seek thy simple music shall be mine;
> I'll strive to win its graces to my lyre,
> And make my plaintive lays enchant like thine. (Opie, in
> Bloomfield 1824, 137–38)

While Thompson's poem was preoccupied with *what* the harp's tunes had to promise his soul, Opie focuses on *how* this message is received: The locus of the poem has moved from the external world of the "wandering spirits of the wind" to the workings of the speaker's psyche. Instead of merely emulating the harp, Opie turns into one herself: The "trembling strings" of the harp become the "trembling nerves" of her body, and with her pledge to "strive to win its graces to my lyre," she casts herself as one of these "breathing instrument[s]," becoming what Bloomfield's collage pamphlet terms a "sonorous body," reverberating external stimuli into internal states.

Opie's linkage of herself with the harp is not just symbolic but underscores a *physiological* commonality between harp and poet. Whereas Thompson's receptive speaker is merely described as affected, and smitten by angelic sounds, Opie observes a more detailed and concrete process of nervous transmission. Her experience of and as the harp is not that of a monadic self, but the result of the workings of a complex organization with component parts. Like the harp, her body is a system of wires ("nerves") triggered into action ("trembling") by an impulse (awoken by "sounds") that follows natural laws ("regular progression") and is controlled by a somewhat autonomous, capitalized "Fancy"—"the faculty by which the mind forms images or representations of things at pleasure" (Webster 1828a; see also Nicholson 1809, 287).

Still, this poet from a prominent family of physicians[4] was not the only one to read such anatomical connotations in the wind harp. Be it John Agg's poem "Ocean Harp," which speaks of prismatic "scenes . . . refracted o'er my brain" and the stringed "Arachne of [the] brain" (1819, 60),[5] Felicia Hemans's harped "shadows of the brain" (1836, 243), or James Montgomery's praise of

---

[4] Opie's father, Sir James Alderson, and his brother John were both physicians and avid essayists. Sir James, for instance, published a treatise on the nature of apparitions, citing natural, mental reasons for such sightings (Society for the Diffusion of Useful Knowledge 1842, 780).

[5] Agg's book is dedicated to the memory of "John Syng Dorsey, M.D.," author of the first manual of surgery published in the United States (*Elements of Surgery*) and the nephew of the "father of American surgery," Philip Syng Physick.

"the harp of thought" awakened by the "passing wind" to the "music of the mind" (1824, 693), the turn of the nineteenth century suddenly found one of its central poetical symbols vibrating to a much more cerebral tune.

Not coincidentally, physicians too began framing the brain in Aeolian terms: A small commissure in the fornix (an arched structure between the hippocampus and the hypothalamus) was from then on to be known as "psalterium" or "Lyra Davidis" (Hyrtl 1846, 17; Cooper 1832, 215; Bell 1802, 49)[6]—both references to King David suspending a harp above his head to be played by the midnightly north wind, so that the tunes might inspire his mind (see also Schwartz 2004, 396; Sachs 2006, 262). The Talmudic antiquity of the wind harp, so often referenced by Bloomfield and other Aeolian poets of the day, was now anatomized into the "Harp of the Brain" (Pierer 1828, 138), a structure of then-unknown function seemingly wrapping itself around the center of the brain (see "H" in Figure 5; Bell 1802, 49), a region hypothesized by many scholars of the seventeenth and eighteenth centuries as hosting the soul.[7]

Figure 5 Cross section of the human brain by Charles Bell (lyra marked as "H").

---

[6] While some anatomical works reference this structure by the term "lyra" before 1800, it was not until the turn of the nineteenth century that this lyre was specified to be King David's proto-wind harp, so often alluded to in poems dedicated to the instrument.

[7] Descartes placed the soul in the pineal gland (Lokhorst 2006), while François Gigot de la Peyronie (1678–1747) believed it to be in the corpus callosum, and Anthelme Richerand (1779–1840) argued it was housed in the pons of the brain stem (Cooke 1824, 54). John Locke objected to such notions and instead claimed the soul to be present throughout the gray matter of the cerebral cortex only (Ione 2016, 86). Or for a more poetical historical overview, via Eleanor Anne Porden in 1816: "But where the soul is situate / Is still an object of debate. / Pythagoras, Haller, Galen, Plato, / Yield for its ample space the brain; / While to the *Corpora striata* / Sage Willis would its rage restrain; / And some more sensual minds debase it / To many a mean ignorable part; / Van Helmont, Aristotle, place it / One in

26 WRITING THE BRAIN

To the poet, there was no escaping this lyre—each lyrical utterance was at once harking back to the genesis of poetry on Aeolus's islands as well as the specific neurophysiological genesis in one's own cranial wind harp. While this transposition of the harp from poetic symbol to physiological analogy has received little scholarly commentary, one of its most prominent literary renditions certainly has.

Samuel Taylor Coleridge's "The Eolian Harp" is perhaps the prime specimen of this anatomical strand of Aeolian poetics as well as a fascinating case study of its biomedical and philosophical source materials (see Trower 2009; Leuschner 2000, 97–129). The history of the poem's revisions also reveals that it was initially planned out as an Aeolian poem in the style of the 1700s. In its first draft, written in August 1795, the harp is a mere accessory to the speaker's declaration of love to his "pensive Sara" (Coleridge's future wife, Sara Fricker). The then-brief poem concludes:

> The noiseless gale from yonder bean-field wafts!
> The stilly murder of the far-off Sea
> Tells us of Silence! and behold, my love!
> In the half-closed window we will place the Harp,
> Which by the desultory Breeze caress'd,
> Like some coy maid half willing to be woo'd,
> Utters such sweet upbraidings as, perforce,
> Tempt to repeat the wrong! (Coleridge 1912, 519)

Like an echo of Thompson's "unhappy maid / Who died of love," Coleridge's harp brings a message to the two lovers; it sings to them about passion but has little to tell each about their very own "music of the mind."

Still, around this time (Trower 2012, 9), Coleridge was beginning to intensely engage with the work of philosopher and physician David Hartley (1705–1757)—Coleridge's son would be named after him. After Coleridge

---

the stomach, one the heart. / Wharton, Schellhammer, both opine / 'Tis in the marrow of the spine, / Boutehoe, Lancisi, are secure / It lurks in the great commissure; / Herophilus believes it lies / Snug in the brain's warm cavities; / While Drelincourt will gravely tell 'em, / Its seat is in the Cerebellum. / And German Soemmering dares maintain / 'Tis in the vapour of the brain, / From which his learned friends presume / He thought it but an idle fume. / And some believe, with Matthew Prior, / That from our toes ascending higher, / Thro' every part successive led, / In age it settles in the head. / Some, with Descartes, it takes command / Imperial in the Pineal gland / . . . / Yet none, while living, e'er could find / The skull devoid of sense or mind" (1816, 157–58).

NATURE'S MIND AND MIND'S NATURE   27

returned to the draft poem in 1796, it read quite differently.[8] Besides describing the instrument in more detail (it is now the "simplest Lute / Place'd lengthways in the clasping[9] casement"), the poet also provides an account of how his nerves and mind receive the harp's vibrant notes and what this knowledge tells him about the nature of human cognition:

> And many idle flitting Phantasies
> Traverse my indolent and passive Mind
> As wild, as various, as the random Gales
> That swell or flutter on this subject Lute.
> And what if All of animated Life
> Be but as Instruments diversly fram'd
> That tremble into thought, while thro' them breathes
> One infinite and intellectual Breeze
> . . .
> Thus *God* would be the universal Soul,
> Mechaniz'd matter as th' organic harps
> And each one's Tunes be that, which each calls I. (Coleridge 1912, 520–21)

The yearning soul of Coleridge's earlier draft is still here—but it is now the yearning of a machine, an accumulation of "passive" substance that has to be excited into thought by outside forces. Even the sense of self (that which "calls" itself "I") is now moved into an almost reflexive position, generated *in reaction* to an external "intellectual Breeze" that plays with its constituent "matter." Cognition has thus been moved from the sphere of the soul into the realm of natural law, and instead of an autarkic self, we now find it dependent on "Mechaniz'd matter." This image is, of course, still a dualist imaginary— a physical brain-machine with an ethereal soul pilot—but it begins to slip

---

[8] Shelley Trower, in her penetrating contextualizations of Coleridge's poem, chooses not to focus on these earlier versions, although she also discusses the "mechanized understanding of mind" that she argues characterized the Romantic period (2012, 34). For a manuscript history of "The Eolian Harp," see Cheshire 2001.

[9] The term "clasping" is fascinating in this context, given the technological history of the wind harp. Is Coleridge merely describing a tightly shut, evenly rectangular casing found in most wind lyres of the time, or is he indeed sketching an instrument with a slanted top that gives off the impression of "clasping"? If the latter is the case, then Coleridge is likely describing an actual Bloomfield harp (see Figure 4), since the Suffolk poet is generally credited with having invented this particular style (Goodridge 2002). Before Bloomfield, wind harps were square, not "clasped." And while Bloomfield had not yet opened his professional harp business at the time of Coleridge's second draft, he was almost certainly already producing and distributing the instruments on a smaller scale.

28 WRITING THE BRAIN

into a rather cyborgian understanding of brain and soul as increasingly interconnected.

What comes to the forefront at this moment in the poem's history is the full force of enlightened, post-Cartesian Sensationalism (cf. Descartes 2012, I.66–75; IV.188–207; Edwards 1967, 417) stretching itself into the dawning century of the brain. Encouraged by increasing medical knowledge of the nervous system, scholars from Thomas Willis, William Cullen, and George Berkeley to David Hume, Joseph Priestley, and David Hartley taught a concept of the human mind centered around a rationalist vision of sensual input that permeated Continental medicine and philosophy alike. To them, the mind was an inert, Lockean tabula rasa upon which the senses create cognition—it was a "mere thinking [machine], set in motion by ... impressions," as Benjamin Rush, the first major figure in American psychiatry and a diligent student of the European Sensationalists, put it (1981, 489). To think meant to filter, structure, and reassemble the data provided by the senses following logical laws of association:

> Individually, [the sense organs] convey little information to the mind; but by comparison and combination, the simple and original affection of feelings of the mind are associated and combined to infinity, and administer to the memory and imagination, to taste, reasoning, and moral perception, the passions and affections, and every active power of the soul. (Bell and Bell 1803, 223)

The mind is figured as a soul-hermit, isolated in a dark cave, interacting with the outside world only via a system of ropes and pulleys, and reassembling the information thus gained into thought.[10] Or, to speak with Coleridge, it was an Aeolian harp, automatically producing harmonious music from random external impulses playing on its strings (cf. Trower 2009, 5–9).

Most of the related medical debates of the late 1700s thus centered around the exact nature of nerves and the manner of nervous transmission: Some espoused notions of "animal spirits" traversing hollow nerve channels;[11]

[10] As Benjamin Rush put it: "On what part of the [human] system do [stimuli] act? On the muscular fibres they produce contraction or motion, on the nerves they produce sensation, and in the brain they produce perception, judgement, and reason; or in other words *thought*" (1981, 181).

[11] See, for instance, David Hume's account of memory: "As the mind is endow'd with a power of exciting any idea it pleases; whenever it dispatches the spirits into that region of the brain, in which the idea is place'd; these spirits always excite the idea, and when they run precisely into the proper traces, and rummage that cell, which belongs to the idea" (1878, 365).

NATURE'S MIND AND MIND'S NATURE   29

Newton proposed "fine ether"; still others, like David Hartley, argued for solid nerves that vibrated when transmitting information (cf. Rush 1981, 224). Still, while increased anatomical knowledge would aid these discussions, it also began to pose problems to such nerve-centric concepts: The more that became slowly known about the brain,[12] the harder it was to conceptually relegate it to being a mere bundle of wires awaiting stimuli.[13]

Mind-Strings

As counterintuitive as it might seem to a modern reader, the brain itself played a surprisingly minor role in debates on the nature of the mind in the eighteenth century. Much of this was due to very practical reasons: Not much was known about it. "Before [the brain] we are much in the dark," Benjamin Rush states around 1795, forcing him to "commit the investigation of the brain to time; *dies doceat*" (1981, 216–17). Even David Hartley himself laments the "unknown internal structure" of this "the great instrument of sense and thought" (1810, 373). In practical terms, the brain at the close of the 1700s was still the brain of Descartes (Figure 6)—a mass of matter in which all nerves terminate, a mere *sensorium commune* that furnishes a soul (housed somewhere within or without) with passions, pleasures, thoughts, and fears generated from the association of stimuli.[14] The 1797 *Encyclopædia Britannica* summarizes this lack of knowledge thusly: "BRAIN, in anatomy, is that large, soft, whitish mass, inclosed in the cranium or skull; wherein all

---

[12] Enlightenment optimism had provided the first wave of psychiatric physicians in the West with a plethora of brain specimens from its newly founded asylums in Europe and (slightly later) the United States (Porter 1995)—and they offered ample evidence of the importance of the organ for the mind. For instance, Johann Ernest Greding 1798's dissections of the brains of 219 asylum patients, which aimed to provide statistical data on correlations between brain malformations and certain psychiatric conditions, the results of which were debated in most psychiatric texts at the turn of the nineteenth century (e.g., in Pinel 1806 and Rush 1812). An English translation of Greding's work was first published in 1798.

[13] For the latter, see, for instance, this conflation of memory and sensory input published in an 1803 article: "Should we suppose an animal which had no brain, but possessing an exterior of great sensibility and extension; an eye, for example, of which the retina was as extensive as that of the brain, and had the property of retaining, for a long space, the impressions it might receive; it is certain, that the animal so endowed would see at the same time not only the present objects; but also those it had seen before" (Leclerc 1792, 21). Here, the brain is a sensory echo chamber with, for instance, the memory of a sound being that very same sound still vibrating (ever so minutely) somewhere deep within the lobed tissue of one's skull.

[14] "The history of the soul's organ [sensorium commune] stretches from the middle of the seventeenth to the close of the eighteenth century. At its beginning stands the philosophical body-soul dualism of Descartes, which, despite numerous challenges, remained the irreducible frame of reference for this entire timespan" (Hagner 2000, 25).

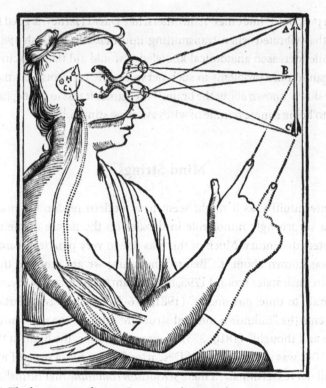

Figure 6 The brain according to Descartes (1687, 79).

the organs of sense terminate, and the soul is supposed principally to reside" (MacFarquhar and Gleig 1797, 511). This single sentence is all this central document of the British mind had to utter about the brain. This idea was indeed quite "Aeolian" in a sense: It understood the mind as long strings extending around a functionally hollow center.

This not only opened up Associationist physiology to scathing philosophical rebuttals by critics like Thomas Reid (see esp. 1785, 99–150, 452–88) but also caused some of the brain-harp's most prominent literary proponents some headache. Coleridge, for instance, would never republish his "Eolian Harp" in its more drastically Hartleyian form. When it appeared again in print in 1803, all references to mechanization had vanished, that "which each calls I" was no longer questioned, and the "intellectual Breeze" had turned into the more mundanely spiritual "Soul *of each*, and God of all" (Coleridge 1912, 10, emphasis mine). Whether this was an early sign of his break with

NATURE'S MIND AND MIND'S NATURE   31

Sensationalism (cf. Richardson 2001, 8–12) or whether the memory of Joseph Priestley being run out of England for his radical materialism was just too fresh in Coleridge's mind is debatable.

Whatever the reason for these significant changes, Coleridge was certainly beginning to question nerve-centric concepts of the mind. In the margins of his edition of *Critique of Pure Reason* he comments in 1801:

> The mind does not resemble an Eolian Harp . . . , conceive as many tunes mechanized in it as you like—but rather, as far as Objects are concerned, a violin, or other instrument of few strings yet vast compass, played on by a musician of Genius. (1992, 248)

Coleridge's comment underscores some of the major issues he was beginning to sense in Hartley and his fellow Sensationalists: Their simplistic notion of a "mechanized" mind that equated the exchange of soul and world with mere physics (vibrations), the seeming inability of Associationism to account for the complexity of thought ("few strings, yet vast compass"), and the lack of original, creative energy in its dogmatic stance on the brain as passive, always reactive matter and a tabula rasa (cf. Richardson 2001, 8–12).

To a poet like Coleridge these were not minor points: Truly embracing the idea of mind as "Mechaniz'd matter" meant conceptualizing the poet as an automaton. Now, all that makes him or her stand out is solely the result of having been "diversly fram'd" by exterior forces. Such a concept leaves very little room for original creative energies or anything truly resembling a "Romantic" imagination. Instead, the poet becomes a *tool*, merely responding to the outside world according to his or her physiological constructedness. Even to a pre-digital mind, the threat of thus being potentially replaceable by a well-designed (perhaps *better*-designed) mechanism was not at all far-fetched.

Robert Bloomfield's son, for instance, would propose just that. Following in his father's footsteps, he fused musical theory with the very question of poetry as such. In an essay titled "The Mechanics of Music," published as a multipart letter to the editor of a musical magazine, the younger Bloomfield lobbied for the creation of lyrical instrument-machines that would perform and perhaps compose music independently of human interference. Music, poetry, and mathematics, the watchmaker therein claims, work on identical principles of harmony, structure, and symmetry and thus all become

32 WRITING THE BRAIN

theoretically executable by apparats.[15] The younger Bloomfield must have been aware of Charles Babbage, who had been conceptualizing his "thinking machines" (as Ada Lovelace termed them) just a few miles down the road.[16] The Aeolian harp, as the first "mechanical" music machine to mimic the brain, had apparently already yielded enough "information on the human information machine" that now "its replacement by mechanics [could] begin" (Kittler 1999, 189). So when this poet-son claims that "all actions that can be performed by the human frame on artificial instruments might be perfectly imitated by machinery" (Bloomfield 1844, 293), might the pen not be one such "instrument"? And could the poets who so eagerly tried to imitate the wind harp then not be "replaced" by a comparable poet-machine?

To Coleridge, this would certainly have been a preposterous idea— but it also underscores the fundamental flaws of Hartley and his fellow Sensationalists and their many problematic presuppositions that critics so loved to prod. Still, Coleridge never quite let go of the idea of "organic Harps" (he rephrased and republished the poem for years to come) and as late as 1825 still spoke of the "Eolian harp of [his] brain" when conversing with friends (Coleridge 1992, 1110).[17] So while he might have come to object to Hartleyian theories, there was something about imagining thought in strictly physiological terms (as an interplay between outer and inner nature in the organ of the brain) that stuck with Coleridge. To a cerebrally inclined poet, the nature of the brain truly posed a problem: If thought were a physiological reaction by a soul, housed in inert cerebral tissue, to external stimuli—as Hartley essentially proposed—where does this leave the subject? How can passive matter constitute an "I" that separates itself from and acts on nature?

A poem like the pseudonymously published "The Spectre Harp" (1829)[18] by fellow English poet "Sforza" provides some first hints at the potentially

---

[15] "Mathematics, music, and poetry . . . are all composed of order and novelty, or of symmetry and variety, which are either blended or alternately exhibited" (Bloomfield 1844, 300). To Robert Henry Bloomfield, the differences between mathematics and poetry are by no means categorical, but function on a continuum of abstraction: "Poetry and mathematics differ in the fact that the subjects or materials of which they are composed are familiar, and those of mathematics remote" (1844, 294).

[16] Indeed, Babbage and the younger Bloomfield very likely met in person around this time. In his autobiography, Babbage describes being potentially swindled by a clockmaker from Clerkenwell. Regretting his decision to lend him money, Babbage then visits all men of said profession in that town (1864, 243)—one of which must then have been our Clerkenwell clockmaker, Bloomfield.

[17] Another famous instance of Coleridge likening the mind to a wind harp is of course his 1802 "Dejection: An Ode," which has the speaker refer to himself as "this Æolian lute" (1912, 1076). For an analysis of the poem in neurophysiological, Hartleyian terms, see Trower 2009, 20–29.

[18] Besides its publication in book form (which received a lengthy but ultimately dismissive review in the Athenaeum of May 13, 1829), this poem was also previously printed in London's Literary

NATURE'S MIND AND MIND'S NATURE 33

devastating lessons such a "dead" brain had in store for Romantics. "The Spectre Harp" poetizes an act of metacognition, of the speaker pacing through his or her own sleeping mind to engage with its nature. The mind at rest here is akin to a "desert-palace" marked by the absence of "the dead enchanter"[19] that used to roam its "lonely chambers." Though tinted in gloom, these halls beckon the speaker "with a nerveless bosom" to further "penetrate the secrets of his dwelling," which attract this Self with "dubious sounds" seemingly designed to "tempt a mind" (Sforza 1829, 91). Hoping to discover "Some new Elysium," this particular Romantic pushes onward and does indeed discover a marvelously "pictured Paradise" stretching out before the mind's eye (Sforza 1829, 93).

Still, these vistas are tinged with the foul odor of decay and upon further examination appear more like "Eden's cemetery" (Sforza 1829, 93) than a Romantic utopia: The "passionate visions of [the speaker's] brain" and his "romantic thoughts," we learn, are only sustained through a "sweet, low, uncertain, soothing sound" filling these chambers that constantly whispers like a "soft voice singing" (Sforza 1829, 94). Pursuing these notes to their epicenter, the speaker is startled to discover them to come not from a coy maiden or an angelic being but from an encased, autonomously playing harp that seems to produce quite marvelous tunes even in the absence of the owner of this cranial palace:

> I found the open casement of a chamber,
> Where music's sound invited me to enter
> From a lone harp, that stood untouched within
> Playing a requiem for the enchanter's spirit;
> And oh! so exquisitely did it throb
> So plaintively its moanings did appeal
> To my heart's sympathy for its absent master,
> So well it seemed to feel the strains it breathed,

Chronicle of December 29, 1827. The *Chronicle* was a weekly publication that competed with the *Literary Gazette* from 1819 to 1828, when it was absorbed by the *Athenaeum* (Matoff 2011, 122). The identity of "Sforza" is unknown, but the *Athenaeum* speculates he must be a young man with an infatuation for Byron and Moore ("The Vision of Noureddin [Review]" 1829).

[19] As Rush remarks: "There is an obvious similarity between sleep and death, for they are both characterized by a loss of consciousness and motion.... Another and probably more important connection between sleep and death [is] the fact that sleep makes it possible for the spirit to range widely through different realms" (1981, 390).

34 WRITING THE BRAIN

> So sensibly, so tenderly it sighed
> Over the memory of departed hours. (Sforza 1829, 94–95)

At the center of the physical mind, the speaker discovers a harp machine playing a sad farewell to the Subject: "memory," "sympathy," feeling, sensibility, and other heart-"throb[s]" of the Soul are suddenly all well within this instrument's tool kit. This organic, pulsating, moaning thing has produced all of the "romantic thoughts" and illusions the speaker encountered in this "desert palace" of the mind—and it needs no helping hand from any magic enchanter.

Once the speaker moves closer to reconcile with this odd entity and makes "an effort to embrace it," this ghastly brain harp fights back (Sforza 1829, 95):

> . . . instantly,
> Changing itself into a skeleton,
> It sprang upon me with a bony grasp—
> A sickly shuddering came upon my frame,
> The blood ran coldly rippling down my veins,
> Till, in convulsive struggles, I awoke,
> With an appalling shriek of agony,
> And found—I had been dreaming. (Sforza 1829, 95)

The thought of such sepulchral materiality of the brain comes as an epileptic fit—the ultimate show of dominance of the brain over the Self. And the "shriek of agony" carried over from this harsh anatomy lesson might as well signify a double sense of "awaking": both from the surreal dreamscape of a brain reflecting on itself and from the Romantic fantasies of a "new Elysium" of the mind.

Such creeping determinism appears coded into the very concept of an Aeolian mind. As such, it was, of course, not new to the English Romantics. Its gloomy prospect lurks behind Coleridge's gradual move away from Hartleyianism. Even his Aeolian acquaintance Shelley[20] seems quite aware of this issue and often effectively exploits the seeming paradox of dead matter thinking without resolving it (or, perhaps, realizing its unresolvability). In his Davidian "Ode to the West Wind" (Wiskind-Elper 2012, 197), for instance,

---

[20] Shelley's poem "To Coleridge" famously begins with the invocation of the air as "spirits of the air, And genii of the evening breeze," and casts Coleridge as "hold[ing] commune" with them (Shelley 1905, 63).

the speaker is quite literally "inspired": His dead mind is filled with the foreign breath of Aeolus[21] to such a degree that very little logical space is left for any non-wind mind to inhabit. It is perhaps the prime example of what Pia-Elisabeth Leuschner aptly frames as the "usurpation of *res cogitans* by *res extensa*" in Aeolian poetics (2000, 97, translation mine). Here, the speaker concludes, still addressing the wind:

> Make me thy lyre, even as the forest is:
> What if my leaves are falling like its own!
> The tumult of thy mighty harmonies
>
> Will take from both a deep, autumnal tone,
> Sweet though in sadness. Be thou, Spirit fierce,
> My spirit! Be thou me, impetuous one!
>
> Drive my dead thoughts over the universe
> Like withered leaves to quicken a new birth!
>
> And, by the incantation of this verse,
> Scatter, as from an unextinguished hearth
> Ashes and sparks, my words among mankind!
>
> Be through my lips to unawakened Earth
> The trumpet of a prophecy! O Wind,
> If Winter comes, can Spring be far behind? (Shelley 1905, 108)

With every notion of a stable subject already having received quite a severe beating in the previous stanzas (the few references to an "I" are couched in hypotheticals or ornamented with violence),[22] this final moment of the poem has the speaker fully become an instrument.

Shelley here clearly takes up the anti-materialist charge that inert, "dead matter"—the "Spectre Harp" that is the brain—cannot possibly produce thought (cf. Wolfe 2016, 9; Yolton 1984, 15; Lawrence 1822, 96). The mind of the speaker sends out "dead thoughts" not merely to fit the piece's metaphors

---

[21] It was, of course, Aeolus's west wind that carried a stranded Odysseus to safety: "The west wind which was fair for us did he [Aeolus] alone let blow as it chose" (Homer 2007, 118).

[22] All of the instances of "I" in the poem are: "If I were" (repeated twice), "I would," "I fall," and "I bleed."

36   WRITING THE BRAIN

of rebirth but because they are expressions of a brain reacting to stimuli—of matter pulsating to the tune of the wind. The poet's task here is like the forest's: He renders the *"harmonics of the harmonics"* of the Aeolian winds (Bloomfield 1824, 107) into physical products of a lesser harmony; the "mighty harmonies" of nature infuse the "leaves" that drop from melancholy trees or the desks of autumnal poets. The "Spirit" in Shelley's skull is quite literally just that: It is "breath," an "animating or vital principle" bestowed by an external nature (cf. Harper n.d., s.v. "spirit"). Spirit or mind, in Shelley, is an outside force, beckoning to be let in. When Paul de Man, somewhat acerbically, charges Romanticism with making "illusionary priority of a subject that had to borrow from the outside world a temporal stability which it lacked within itself" (1971, 200), he might well be describing this notion (while nonetheless choosing to ignore the Romantics' acute self-awareness of an "I" in flux).

Still, in imaging his speaker as an Aeolian android with wind entering his skull and leaving it "through [the] lips" of the "clasping casement" (Coleridge again) of his cranium, Shelley is describing a closed system or loop. Here, to think and feel becomes a mere passing on of impulses, a "Begeisterung," as Hoffman puts it.[23] Nature plays on the brain of the poet, so his utterance can in turn reinvigorate a physical universe once more rendered passive ("unawakened Earth"). But in Shelley's piece, both nature (*res extensa*) and mind/spirit (*res cogitans*) are congruous; both are *matter*, and as such, they have the potential to be invigorated into a state of "mind." Where Coleridge was sensing a denigration of thought into mechanics in the proto-neuroscience of his youth, Shelley seemed to discover a quite different promise in the very same concepts: that of an "awakened Earth"—of Nature itself becoming mind (or at the very least, realizing its mindedness). For every dead "Winter" of passive matter, there will be a "Spring" of mind.

"Ode to the West Wind" thus almost reads like a reminder to Coleridge of one of the central tenets of Romantic Aeolianism. Much like the wind harp is said to not merely *produce* sound but instead pluck it from the air ("air prism"), the brain-harp does not create mind (as almost an afterthought) but *finds mind in matter* and makes it speak. Still, while this thought certainly is "Sweet though in sadness," it leaves the central conundrum unresolved: If Shelley's poem is Nature pressed through lips, how can he speak of

---

[23] "[Nathanael] went as far as claiming that it is foolish to believe that art and science is produced from one's own self-actuating whims; for the enthusiasm required for creation does not come from within but is an inspiration from a higher principle located without" (Hoffmann 2008, 28, translation mine). The German word *Begeisterung* ("enthusiasm") literally means "to fill with mind/spirit."

"my words"? If thought is an *aspect* of matter, how could one conceptualize "self" accordingly?

"For whatever matter be, . . . mind is nothing more than a modification of it," the materialist theologian and vibratory Associationist Joseph Priestley (1733–1804) had already claimed in the 1770s (1777, 355). Still, "whatever matter may be" was rather unclear with regard to the brain: Until the middle of the nineteenth century, no coherent medical theory on *how exactly* the mattered brain might function stepped up to supplant increasingly questionable Hartleyian speculations about an organ composed solely of oscillating "vibratiuncles."[24] With cranial neurons (and like concepts) still unknown, the "whitish mass" (MacFarquhar and Gleig 1797, 511) behind one's eyes invited speculation. At the same time, the common Sensationalist definition of matter—"something that is extended, figured and coloured" and can thus be perceived as a *thing*—was losing its usefulness because of its seeming inability to account for the phenomenon of matter becoming mind in the brain (Knowlton 1829, 426). Where physician-philosophers made initial, often eerily suggestive, Aeolian arguments about the embodied (enmattered) nature of the mind, another group of hyphenated thinkers had already begun to supply one possible means to this process—and this time, they were chemist-philosophers.

Mind-Matter

Although the "Romantic" nature of early-1800s chemistry, especially by example of the *Naturphilosphie*-style writings of electrochemist Humphrey Davy (1778–1829), and its connections to poetry has received some scholarly commentary,[25] it is the concept of matter proposed by John Dalton

---

[24] Scottish poet-philosopher and physician Thomas Brown (1778–1820), for instance, complained that "it has always seemed to [him] peculiarly wonderful that such a hypothesis [like Hartley's] should have been formed by a physician, to whom the structure of the brain and its appendages must have been familiar" and states that by the early 1800s, Hartley's "system of *vibrations* and *vibratiuncles* . . . [had] fallen into merited disrepute" (1827, 320).

[25] See Levere 1977; Danby 2000; Chaouli 2002; Ruston 2013; Cunningham and Jardine 1990, esp. 213–40, 263–79, 295–307. For a more nuanced overview of the chemistry of the 1800s, see Knight 2016, 59–118. Davy himself appeared to have followed Associationist concepts of the mind. In an essay on the nature of vision, he writes that once stimuli are transmitted by nerves to the brain, "in the brain . . . correspondent ideas are associated together according to certain laws." Instead of Hartleyian vibrations, the chemist seems to subscribe to the ether theory of nervous transmission: "Is it not probable that the existence of some fine ethereal principle in the brain and nerves is the immediate cause of sensible or perceptive action?" (1839, 82).

38 WRITING THE BRAIN

(1766–1844)[26] that remains an underappreciated, though crucial, pillar to the Romantic construct of matter and mind: atomic theory. This theory suggested a novel way of combining seemingly determinist mechanics with, in the long run, a mind-boggling sense of scale and a deep appreciation for the *active* properties of matter. Indeed, while from today's perspective "atomism" might appear counterintuitive to the sweeping vistas and generalizing claims often associated with Romantic thought, atomic chemistry did not relish a fetishizing of minutiae but initially understood itself as a universalizing concept, a "synthetic" vision of life, as Dalton himself calls it:

> When we attempt to conceive the *number* of particles in an atmosphere, it is somewhat like attempting to conceive the number of stars in the universe; we are confounded with the thought. But if we limit the subject, by taking a given volume of any gas, we seem persuaded that, let the divisions ever be so minute, the number of particles must be finite; just as in a given space of the universe, the number of stars and planets cannot be infinite. . . . All the changes we can produce [upon matter], consist in separating particles that are in a state of cohesion or combination, and joining those that were previously at a distance.
>
> . . . No new creation or destruction of matter is within the reach of chemical agency. We might as well attempt to introduce a new planet into the solar system, or annihilate one already in existence, as to create or destroy a particle of hydrogen. (1808, 212)

Individual existence, under the gaze of atomic chemists, had become a momentary cohesion of particles from a loose, universal fabric. Each being could now find the universe itself mirrored in its own atomic makeup. Instead of dead matter passively played on by the airy fingers of Nature, *res cogitans* and *res extensa* had now become not only interpermeable but interchangeable. Or as Schelling put it: "Between mind/soul [*Geist*] and matter, you can discern such a multitude of intermediary states [*Zwischenmaterien*], which become ever finer and finer until we reach a point where spirit and matter

---

[26] One exception is Levere 1977. Of course, the notion of the atom is an ancient one (cf. Pullman 2001). Still, Dalton is generally credited with introducing it to and rearticulating it for modern chemistry (Burdge 2016, 5). On the debates over his atomic theory (and the harsh backlash it faced), see William Hodson Brock's excellent *The Atomic Debates*. "Had there been a closer liaison between physics and chemistry," Brock therein rightly observes, "there might have been less skepticism towards the atomic theory" (1967, 59).

are one" (in Hagner 2000, 154, translation mine). Perhaps it was not the brain as a whole that ought to be conceptualized as a harp, but each of its minute particles, rendering the mind a vast orchestra, not a soloist.

This psychophysical reading of atomic theory was by no means a misreading of science by philosophers—and some chemists were surprisingly direct about it. The eloquent claims of a somewhat obscure French-Canadian surgeon by the name of François Blanchet (1776–1830)[27] might set the stage for us. While his major work, *Recherches sur la médecine, ou l'application de la chimie à la medécine* (published in New York while Blanchet studied at Columbia College), was never fully translated into English, it quickly made its way into the anatomical literature of England and the United States. The American *Medical Repository* of 1808, for instance, renders Blanchet's introductory paean to "la Chimie" as follows:

> By [Chemistry's] direction, heterogeneous atoms could arrange themselves into filaments, and filaments bend to circles and, as filament placed itself beside filament, and circle associated itself with circle, muscles and vessels, and by easy process, were constructed. To these, by a necessary consequence, *irritability*, or a *susceptibility* of *impression*, inhered. If, proceeding from this *simple* form, she chose to be a little more *complicated*, she could combine other atoms into brain and nerves, and impart to them *sensitive* and *voluntary* power. . . . Lastly, out of this combination of elements, *mind* would arise as one of their modifications, and *intellect*, and *memory* and *passion*. (In Mitchill and Miller 1808, 172)

Instead of dead *stuff* transmitting impulses, atomic matter had now taken on properties long held to be fundamentally opposed to it: irritability, sensitivity, and voluntary power. What would startle the speaker of the "The Spectre Harp" into determinist panic exudes in Blanchet a peculiar beauty of its own: Instead of a perfect reactive automaton generating a singular mind, mind oscillated as a multifold, atomic property.

Something abstract as "thought" could thus be seen as a complex harmony of physics and matter. The *Dictionary of Science, Literature, and Art* by fellow chemist William Thomas Brande (1788–1866) seems to agree with such notions and hence defines "matter" explicitly running counter to Sensationalist lines, claiming that "of the intimate nature of [matter] the

---

[27] On Blanchet, see Castonguay 1996.

40   WRITING THE BRAIN

human faculties cannot take cognizance . . . by observation." Given the divisible, atomic nature of matter, it is thus logical to conclude that "numbers of atoms of one kind [can also] admit of combination with some determinate number of another kind . . . and thereby [form] compound atoms, [that have] properties peculiar to that combination and different from the known properties of their elemental atoms" (Brande 1853, 727).[28] Cognizance ceases to be a mere mechanical, reactive process: Instead of passive matter, rattling sensory data through a rational mechanism to associate them into thought, atoms now moved atoms, and particulars formed complexes that had properties categorically different from those of their building blocks.[29]

While established chemists like Dalton or Brande were certainly careful not to overinterpret these notions with regards to the nature of the mind and the self, philosophers and literary figures were less hindered by professional decorum, and none perhaps as little as the certainly eccentric, perhaps insane,[30] John "Walking" Stewart (1747–1822). A close personal and familial relation to Brande (who would edit and publish Stewart's posthumous work), this peculiar materialist-idealist philosopher and self-styled modern Pythagoras (as well as a mentor of sorts to Wordsworth)[31] is most frequently

---

[28] These ideas were certainly on far ends of the scientific discourse and debated quite controversially at the turn of the nineteenth century: "An atom [is] essentially inactive, and destitute of feeling or thought," wrote an Irish theologian by the name of William Bruce in the 1820s, echoing the famous charge against such claims and repeating the long-established truism that "matter is essentially inert and unintelligent; therefore, what is active and intelligent cannot be matter." Since "matter is not essentially active," it may also not "become so by arrangement and organization of parts," according to Bruce (in Ely 1822, 287–88). This point of contention, of course, has its roots in the eighteenth century, with fervent debates between deist freethinkers and more orthodox theologians/philosophers over whether certain active forces (such as gravity, repulsion, etc.) were *inherent* to matter or implanted from without by a god (see Yolton 1984, esp. 29–48, 90–106).

[29] When ecocritic Onno Oerlemans thus critiques the Romantics as, ultimately, relying on a theistic idealism that necessitates a God as a deep system of mindedness—a concept that is "always only a faith, a belief" (2004, 43)—he unfortunately overlooks this particular brand of atomism. Not surprisingly, Oerlemans's use of "matter" rehearses notions of inertia and "deadness." With this context in mind, one might be tempted to question his view of the Romantics as hypothesizing that "spirit exists first, matter second" (2004, 34) and that the latter is "essentially or at least initially inorganic, the substance and substantiality of the world" (2004, 34) and likened to "death" (2004, 42).

[30] Stewart himself concludes—characteristically humbly—that based on his writings, it might be determined either "that [his] mind is insane, or that it is a just exemplar of its own discoveries, in carrying sensation and perceptions far beyond the powers of its species." Still, he must "candidly avow, however unfavorable it may be to [his] public appeal for [his] reputation of sense or insanity, that [he] never met an individual who could comprehend" his ideas (Stewart 1810, 82).

[31] On the relationship between Wordsworth and Stewart, see the work of Kelly Grovier, especially 2005 and 2007, as well as Gravil 2015, 177–201. For more on Stewart as a philosopher, see also Claeys 2014 and Jones 2015. For a brief overview of his life and thinking, see Heisler 2003. Stewart himself also includes a brief autobiography of sorts with his *Opus Maximum* (1803, xiii–xxix). See also Thomas de Quincey's favorable portrait of Stewart (1876, 595–602).

NATURE'S MIND AND MIND'S NATURE   41

encountered in scholarship as a colorful footnote. But, as Brande's involvement suggests, he is in many ways a figure central to Romantic theory.

In his work, Stewart initially played with a mechanistic, Aeolian vision of the mind[32] before breaking into Romantic atomism by the early 1800s. In the same year that Dalton was first presenting his atomic theory (1803), and incited by the violence of the French Revolution, Stewart put out his book-length essay *Opus Magnum*[33] (and later, in 1810, its follow-up *The Philosophy of Human Society*). The book constituted his attempt at moving human thinking itself from "contingency to system" to resolve the pressing matter of his day. Instead of each person acting in self-interest (or even as selves), Stewart here proposes what he considers to be a novel way of seeing the world as a holistic, interconnected system of active matter: a perspective on life he terms *omoousia* (one being).[34]

Only thinking of self, matter, and world as fundamentally one and the same, Stewart argues, can enable humanity to progress toward justice, good, and happiness. Dedicated to the American republic, the essay frequently lapses into semipoetical catalogs to celebrate the chemical nature of existence as such—of atomic matter transcending individual life and permeating all:

[32] In his 1790 *The Apocalypse of Nature*, for instance (republished in 1837 under a different title), Stewart speaks of the self as "impressions made upon the animal by means of the senses or corporeal nerves, [that] terminate in the brain, and form what is called the mind" and considers "vibrations" one of the primary means of "the universal inter-revolution of matter" (1837, 28). A prospectus for a 1796 lecture in Philadelphia on concepts of mind suggests likewise (see Stewart 1796).

[33] The exact timeline of Stewart's writing of his "Great Essay" is unclear. Still, given his nomadic lifestyle, the apparent hastiness of the piece's composition (some passages toward the end are barely paraphrased copies of earlier ones), and its direct political contexts (esp. the redeclaration of war between the United Kingdom and France in May 1803 over the Napoleonic occupation of parts of the Netherlands), it is unlikely that Stewart began much earlier than 1803 to compose the piece. It is striking that the book shares a title with Coleridge's collected philosophical musings—they are the only two print publications on file calling themselves an "Opus Maximum" in the nineteenth century (according to the WorldCat database).

[34] While Stewart was frequently charged with atheism, and was an outspoken opponent of what he considered "mysticism" and "superstition" (see, e.g., Stewart 1803, 183), his writings offer themselves as more of a corrective than a flat-out rebuttal of Christianity and religion as such. The term "omoousia" is obviously a reference to the First Council of Nicaea (AD 325) that debated the Trinitarian nature of Jesus. The term was used by Athanasius of Alexandria (c. AD 296–98) to suggest that God and Son are one of the same (as opposed to "omoi usia"—one *like* the other) and was employed to discredit the so-called Arian heresy, which refused to subscribe to such a notion. Stewart's work is certainly engaged in core Christian concepts that originate from Nicaea, such as consubstantiation and transmutation, but renegotiates them through atomism and natural science. For more on the term "omoousia" in its Christian context, see Bethune-Baker 1901.

42 WRITING THE BRAIN

The improvement of chemical knowledge I have myself advanced without the use of instruments; or even the common professional knowledge of the art, by attention and contemplation alone. I observed the various phenomena of miners, working in a copper mine, absorbing the particles of the ore, both with their bodies and their clothes. I observed, that whenever sickly bodies came in contact, in their atmospheres only, with healthy bodies, contagion of disease has ensued to the healthy. I have observed, that all bodies sending forth such strong and pungent effluvia have been absorbed by bodies in their vicinity, proved by musk, which penetrates the corks, and gives savour to wine in bottles. Upon these phenomena I have established the knowledge of omoousia, or union of all being, which identifies the interest of matter in time and futurity. (Stewart 1803, 172–73)

More than half a century before Walt Whitman would compose his love poems to the incessant composting of nature's atoms,[35] "Walking" Stewart already attempted to glean a strikingly similar philosophy from the chemistry burgeoning in his day[36]—or, as he framed it, he intuited said chemistry while contemplating nature with an omoousian gaze.

Stewart's chemical belief truly is an explicitly *atomist* one, subscribing to the extreme interpretive ends of the universalist claims Dalton was putting forth at the same time:

Man rises in formation or birth, through the generative power of parents, from the great mass of matter which surrounds him, into which he again dissolves by death, to which he in substance but not identity, is again returned. These truths are discoverable in the phenomena of contagion, both of living and dead bodies, and in the universal circulation or interchange of matter, between all bodies whose atmospheres come in

---

[35] See esp. his "This Compost," initially published as "Poem of Wonder at the Resurrection of the Wheat" in 1856, which bears the mark of atomic chemist Justus Liebig, whose work Whitman reviewed in 1847 (Schöberlein 2013, 62–63). Liebig, very much a dualist when it came to the embodied mind, was nonetheless partial to Aeolus's instrument: "Like the harp sings, when its chords are moved by the wind, so the brain thinks through (in?) the body's metabolism" (cited in Bischoff 1874, 93). Translation mine; for a contemporary translation, see Liebig 1859, 293.

[36] In *The Philosophy of Human Society*, Stewart declares: "My muse shall be the science of animal chemistry. . . . The laws of animal life, and those of chemical powers and affinities of matter, explain to man the all-important knowledge of self and nature" (Stewart 1810, 4).

contact: such phenomena demonstrate the union of all matter in essence and interest.

The man riding or driving a horse incorporates his particles or atoms, reciprocally, with the body of the horse. (Stewart 1803, 138)

A human, in this system, constitutes what Stewart would later term a "single mode of agency" within a "universal patiency" of circulating matter (Stewart 1810, 54). The active properties of a person—his or her mind, thoughts, impulses—are thus attributes of matter temporarily arranged in such a way as to produce a sense of self and coherence. Here, the brain becomes a node that gives form to mind, a concert hall to its harpists. "Thought or mind is nothing but the substance of the brain in action," Stewart therefore claims, "though the understanding of man is incapable of discovering by what subtle process matter modifies itself into thought" (1803, 55). While the brain is able to contemplate *external* matter, the nature of itself *as matter* is hidden from it. It is like one of Plato's cave dwellers, obsessed with its senses and unable to move beyond their flickering impressions.

This "total incapacity of man to self-examination" is thus the result of what Stewart calls a "defect of the faculty of perception" (1803, 24). Still, to this pre-Hegelian dialectic thinker (Heisler 2003, 18–20; Stokes 2015) this does not mean that oomouisa is completely out of reach. With a little willpower, the "defect" can perhaps be not fixed but at least ameliorated. Instead of being lost to phantasms of perception and festering superstitions, it is possible, Stewart argues, to acquire knowledge about the fundamental atomic interconnectedness of the universe via a process situated somewhere between "ratiocination" and "meditation" (1803, 64–71, 30–34). As the focal and starting point for such inquiries, Stewart proposes the question "What is the nature, end, and means of human existence?" (1803, 67). By consciously stripping away memories and experiences tied solely to one's individual context or cultural prejudices and instead scaling up certain personal experiences (for instance, of the "musk, which penetrates the corks, and gives savour to wine") and scientific insights (on the "transmutation of . . . atoms") to the scale of the totality of matter (Stewart 1803, 66), one can gradually employ the brain's faculties to experience partial omoousian insights that can be ratiocinated into a system—all provided one has an open mind and discipline of thought.

44 WRITING THE BRAIN

Mind-Atoms

It is not surprising, then, that one of Stewart's most ardent admirers,[37] fellow wanderer (Shell 2015, 78) and almost-chemist William Wordsworth,[38] would be among the most prominent poetic practitioners of omoousia of his day. While "Walking" Stewart's influence on Wordsworth's work has been traced into several works (Stokes 2015; Grovier 2007, 2005), they are nowhere as visible as in the specific passages where Wordsworth describes the awakening of a mind to poetry. Especially here, it becomes clear that the Romantic self was indeed not as much "imperiled by . . . [the] atomization of nature," as Catherine Rigby suggests, as, instead, this atomization became one of the *means* by which "an inner connection between the transcendent dimension of our humanity and the 'grandeur' of the natural world" could be illustrated (2004, 23).[39] Always connected to walking and wandering, such moments in Wordsworth are triggered by vivid, perhaps too vivid, impressions of nature that necessitate an almost reflexive retreat of the prospective poet-mind into sensory isolation—in crags, caves, and so on—to reflect these moments and ratiocinate them into an appreciation for the fundamental interpermeability of self and nature.

Especially in two pieces written while preoccupied with Stewart's theses—*The Prelude* and *The Excursion* (Parker 1972, 87)—Wordsworth's omoousian indebtedness becomes apparent. In the first book of the former, for instance, the autobiographically tinged mind-harpist borrows a boat "tied to a willow tree / Within a rocky cave" (1889, 240) and paddles

---

[37] Wordsworth likely met Stewart in revolutionary Paris around 1792–94 and was apparently fascinated by him. The odd thinker soon became a frequent topic in the discussions of the poet with his friend Thomas de Quincey, whom he had first met in 1804. In his autobiographical work, de Quincey mentions "pamphlets which [he] received from [Stewart] at Grasmere . . . on the subject of Tyrannicide" and records Wordsworth's impression of them (de Quincey 1886, 131). De Quincey had moved to Grasmere in 1809 and lived there for ten years. A letter to Wordsworth's sister Dorothy in 1809 (Jordan 1962, 176) also indicates that Wordsworth's "friend" (Fairer 2009, 53) Stewart was a familiar name in their circle and his theses could be debated without much introduction. Grovier (2007) has persuasively argued that the figure of the orientalist "phantom drifter" in Wordsworth's *The Prelude*, the autobiographical poem that famously accompanied the poet from 1798 to his death, is a coded homage to Stewart, suggesting a lifelong engagement of Wordsworth's with the person and work of the philosopher who had allegedly walked from India to France.

[38] Wordsworth had been a student of chemistry and apparently considered giving up poetry for the pursuit of the science in early 1801 (Knight 1889, 229).

[39] Of course, Rigby's use of "atomism" is not meant literally here but is employed to characterize what she sees as reductionist science that turns nature into a "mechanistic" object—a mischaracterization, in parts, of at least some of the scientific output of Wordsworth's day (and its poetic echoes).

NATURE'S MIND AND MIND'S NATURE   45

onto a midnightly lake. There, he is utterly startled by the view of a "summit of a craggy ridge" outlined against the starlit sky and appearing as if imbued with "voluntary power instinct" and a "purpose of its own" (Wordsworth 1889, 240). Upon retreating in fear, the young poet-mind begins to make sense of the experience of nature taking on such startling qualities of mind:

> . . . after I had seen
> That spectacle, for many days, my brain
> Worked with a dim and undetermined sense
> Of unknown modes of being; o'er my thoughts
> There hung a darkness, call it solitude
> Or blank desertion. No familiar shapes
> Remained, no pleasant images of trees,
> Of sea or sky, no colours of green fields;
> But huge and mighty forms, that do not live
> Like living men, moved slowly through the mind
> By day, and were a trouble to my dreams.
>
> Wisdom and Spirit of the universe!
> Thou Soul that art the eternity of thought
> That givest to forms and images a breath
> And everlasting motion, not in vain
> By day or star-light thus from my first dawn
> Of childhood didst thou intertwine for me
> The passions that build up our human soul;
> Not with the mean and vulgar works of man,
> But with high objects, with enduring things—
> With life and nature—purifying thus
> The elements of feeling and of thought,
> And sanctifying, by such discipline,
> Both pain and fear . . . (Wordsworth 1889, 240)

Faced with the prospect of nature awakened to "mind," the young poet begins to comprehend himself as merely one mode of being in a vaster network of active matter. All pleasant sensory input falls by the wayside for a moment and all memory is stripped bare as he briefly gazes past appearances (shapes, images, colors) and into essence (form).[40] Realizing, as Stewart did, that "mind is nothing but the substance of the brain in action" (Stewart 1803, 55), the speaker's mind now transforms into that eerie mass towering against a starry night sky that had previously startled him. Now, "o'er [his] thoughts / There hung a darkness" like the "huge peak" he encountered as a "black" mass radiating "voluntary power." An "eternity of thought" is here not a promise to the self, one's soul, or any individual's mind but is the constituent "Wisdom and Spirit of the universe!"

Turning into a black canvas of matter, this young mind momentarily aligns its internal matter—its brain—with the external matter of the universe and discovers them to be one "in essence and interest" (Stewart 1803, 138). The speaker's "human soul" is not a monadic spirit particular to himself but a mere component part of a vaster (now capitalized) "Soul" of the universe. The address "Thou Soul" thus no longer points within but has to be uttered outward. Still, this realization is not the fatalist nightmare it initially seems but an idea that "purif[ies]" and "sanctif[ies]" mindedness itself in a synthetic, dynamic vision of life. Instead of threatening "solitude" or "blank desertion," it reveals thought and nature to be one and the same—though only "in substance but not identity" (Stewart 1803, 138).

It is more than mere ornament, then, that when describing the incessant ruminations of a brain, Wordsworth sees the need to frame such a passage with the entering and exiting of a cave. A semiparadoxical setup for what is ultimately the description of a lake adventure, the passage almost reads as though the speaker has never really left the cave—underscored by the scenario's almost total lack of sensory input, save a few stars inhabiting the all-encompassing blackness of their firmament like glowworms on the ceiling of a grotto. Akin to the desert palace of "The Spectre Harp," stepping into a cave in Wordsworth's world always also means entering one's mind—braving the darkness of the bony "casement" that is the skull and marveling at the complexities found there. In these moments, Wordsworth is performing cranial introspection. In the case of *The Prelude*, the lightless cave harbors not a terrifying, skeletal automaton prone to attacking the speaker but an organic, leafy harp: an impossible willow tree that once held the Aeolian lutes of "Zion's children . . . / When by Euphrates' stream they made their plaint"

---

[40] Coleridge put the Romantics' distinction between the two terms most succinctly in the late 1810s: "Difference of Form as proceeding and Shape as superinduced—the latter either the Death or the imprisonment of the Thing; the former its self-witnessing, and self-effected sphere of agency" (2002, 131). See also Rowland 2012, 250–52.

NATURE'S MIND AND MIND'S NATURE 47

(Thompson again, in Bloomfield 1844, 130).[41] There is a sense of mourning in this symbol for the brain—a hint of the death of an individual cerebrum—but also a promise of growth and regrowth in the universe's much vaster "eternity of thought."

Such an equation of cave and skull becomes even more obvious in Wordsworth's "The Wanderer" (originally a separate poem, but later folded into *The Excursion*). Describing the youth of the eponymous protagonist, we find him experiencing a very similar developmental arc, first embracing mere impressions but later, upon deeper meditation in sensory deprivation, ratiocinating them into a vision of the universe that finds thought in matter and every boundary between mind and nature erased. After listing the many wondrous things this young wanderer (perhaps "walker") beheld in nature, the piece continues:

> ... he thence attained
> An active power to fasten images
> Upon his brain; and on their pictured lines
> Intensely brooded, even till they acquired
> The liveliness of dreams. Nor did he fail,
> While yet a child, with a child's eagerness
> Incessantly to turn his ear and eye
> On all things which the moving seasons brought
> To feed such appetite—nor this alone
> Appeased his yearning:—in the after-day
> Of boyhood, many an hour in caves forlorn,
> And 'mid the hollow depths of naked crags
> He sate, and even in their fixed lineaments,
> Or from the power of a peculiar eye,
> Or by creative feeling overborne,
> Or by predominance of thought oppressed,
> Even in their fixed and steady lineaments
> He traced an ebbing and a flowing mind,
> Expression ever varying! (Wordsworth 1889, 417)

---

[41] Psalm 137:1–3, KJV: "By the rivers of Babylon, there we sat down, yea, we wept, when we remembered Zion. We hanged our harps upon the willows in the midst thereof. For there they that carried us away captive required of us a song." The passage has often been read in an explicitly Aeolian manner, especially in the 1800s (Erhardt-Siebold 1931–32, 359).

48   WRITING THE BRAIN

Like before, this proto-poet's notion of nature is not a picturesque but a digestive one: Nature is something to be distilled, worked through, to reveal its inmost value. This working-through takes place "Upon his brain," a mental process externalized as the speaker musing solitary "in caves forlorn, / And 'mid the hollow depths of naked crags." Such stony enclosures, here, are not a mere place of reflection but themselves become cerebral. The speaker's brain seems to infect their rocky walls, turning "even . . . their fixed lineaments" into signifiers of an "ebbing and a flowing mind" surrounding him. Indeed, "lineament" is a word that specifically points to the brain in Wordsworth, with the first chapter of *The Prelude* already conceptualizing joy as "substantial lineaments / Depicted on the brain."

Certainly, Wordsworth is revealing himself as decidedly un-Hartleyian here,[42] appearing to subscribe to something more along the lines of Humphrey Davy's "fine ethereal principle" ebbing and flowing "thought," sensation ("eye"), and "creative feeling" through "hollow" brain channels (cf. Smith 2007).[43] Performing something close to a hall-of-mirrors effect, the speaker of the poem turns into a single atom of thought in the much vaster brain of nature, while at the same time realizing the elemental, combinatory makeup of his own mind. Here, a brooding mind reflecting on nature morphs into nature, and nature, in turn, can be understood as mind. From the smallest particle within the speaker's skull to the totality of the universe, mindedness reveals itself as a quality of matter with its respective "[e]xpression ever varying" in degree but never entirely dissipating. Hegel scholar Wayne George Deakin reads this moment as a speaker torn between

---

[42] Similar echoes can also be found in Wordsworth's "Cave of Staffa," where the cave turns into the mind of the mythological Fingal, each thought morphing into "shadowy Beings" rummaging through "every cell" of his mind (Wordsworth 1889, 721). Here, it is the "motion" of air that invites the "ghost of Fingal to his tuneful cave / By the breeze" (Wordsworth 1889, 721). His ethereal thoughts run explicitly counter to vibratory Enlightenment concepts like Hartley's, which Wordsworth clearly refers to as "presumptuous thoughts that would assign / Mechanic laws to agency divine" (Wordsworth 1889, 721). Like the young "Wanderer" experiencing the mind of nature rushing through the lineaments of his brain-cave, "Cave of Staffa" harkens back to concepts of "animal spirits" or "fine ethers" blowing through the brain, while also undermining the spirit/nature-divide that characterized many of this theory's earlier proponents. The "desolate harp" (Wordsworth 1889, 721) that is being played at the center of "The Cave of Staffa" is a literal "wind harp": It is bellowing wind pressed through craggy cerebral formations without the necessity of an actual stringed, mechanical instrument. Still, thought/ music here is nonetheless the result of *physical* force, an active aspect of nature revealing mind in landscapes inner and outer. (Of course, the cave might more appropriately be termed a "pipe," a "flute," or an "organ" here—but this would clash both with the tale of Fingal and the mechanistic notion of the mind by then so closely tied to the image of the Aeolian harp.)

[43] The thesis of hollow nerve channels hosting a gaseous medium (ether, air, etc.) is an old one, dating back to Galen, embraced by Hume (see footnote 11) and Newton, and surviving (though with considerable criticism) as late as the mid-nineteenth century, at which point more advanced

two poles, "one more materialistic and one more idealistic and organic" (2015, 85). But if Wordsworth is indeed attempting a Hegelian (Stewartian?) move here—would "organic" (as in "organic chemistry") not be his synthesis? The atomic nature of matter was indeed not antithetical to "materialism" as such but an attempt to fuse a world of "essences" hidden from the senses with a universalist, enmattered worldview.

Coleridge, as the more philosophically inclined of this leading couple of British Romanticism, certainly would have agreed. After his move away from mechanistic Aeolianism, he also found himself enamored of chemistry and attempted to, more theoretically, "ratiocinate" its lessons into an epistemological concept. While his connection to "Walking" Stewart is more tenuous than Wordsworth's—he was very much in the air for Coleridge, but it is unclear if they ever met (see McFarland 1969, 100)[44]—Coleridge was indeed quite familiar with the father of the atomic theory, John Dalton. Around the time Wordsworth first met Stewart, in the last decade of the 1700s, Coleridge made an effort to be included in Dalton's intellectual circle at Manchester. Soon he found himself frequently attending the chemist-philosopher's debates, where he learned of "the late discoveries in chemistry and other sciences" and was likely even present when "Dalton first broached his then undefined atomic theory" around 1793 (Nodal 1877, 161). And it wasn't a brief infatuation, either: The poet still mentions Dalton, positively, in a letter to Wordsworth more than ten years later (in 1804), when the then "Lecturer on Natural Philosophy" was already publishing on his atomic thesis (Coleridge 1895, 457) and Coleridge had begun to distance himself from Harleyian notions of mind.

While Coleridge would come to denounce the "atomistic scheme" as early as 1817 (in Levere 1977, 349), it is important to note that this does not equal a flat-out dismissal of atomic thought. Instead, he may have been critiquing a particular brand of atomism that had by then taken over British and French thought (including the minds of Dalton and Harvey)—and one he considered to be a relapse into a rationalist theory of nature based in passive, inert matter

microscopy rather conclusively demonstrated the "solid" nature of nerve cells (Ford 2007, 31–35). Still, given that Wordsworth frequented the same circles as Davy around 1801, he seems a likely candidate to have provided these ideas (Knight 1889, 229).

[44] Coleridge was familiar, though, with Stewart's relation William T. Brande, the chemist who would publish the Scots philosopher's posthumous work. He had read Brande's *Manual of Chemistry* and listed him among those chemical thinkers whose work was fit to cause an almost inevitable "*enlargement & emancipation* of [a student's] Intellect" (Levere 1977, 364–66, 374).

50   WRITING THE BRAIN

moved by mere physics. The debate Coleridge was taking sides on, historian of chemistry Trevor H. Levere rightly observes, was *not* one "between dynamism[45] and atomism as such, but between dynamical and mechanical, passive atomism" (1977, 360). Indeed, in his theoretical pursuits to uncover the "ultimate identity of laws of mind and laws of nature" (Levere 1977, 360), Coleridge would attempt to reconceptualize the atom as an idealist entity. So, while he might have been frustrated with the direction Dalton's theories had taken by then, the struggling philosopher was not willing to let the atom go so easily.

Especially his 1818 manuscript *Hints Towards the Formation of a More Comprehensive Theory of Life* can be read as Coleridge's defense of the true essence of chemistry against what he considers to be its abstractors— mechanist theorists of the atom eager to turn nature into mere ammunition for mathematics and denigrating matter into objecthood. Giving voice to his philosophical disappointment with the chemist he once admired, Coleridge writes:

> Abstractions are the conditions and only subject of all abstract sciences. Thus the theorist (vide Dalton's Theory), who reduces the chemical process to the positions of atoms, would doubtless thereby render chemistry calculable, but that he commences by destroying the chemical process itself, and substitutes it for a *mote dance* of abstractions. (1848, 51)

Coleridge even goes as far as calling the "atom itself . . . a fiction formed by abstraction" (1848, 52). The essence of matter, its living properties, to this enraged reader of chemistry, appeared to fall by the wayside for a resurgence of such seemingly obsolete Sensationalist measures as weight, mass, and shape. Contemporary atomic chemists, to Coleridge, relished in the "Death or the imprisonment of the Thing" (its outwardness and surface) while obstructing a view of the "self-effected sphere of agency" that he senses at the core of all matter (2002, 131).

Still, even given this rather furious dismissal of atomism, it remains true that, as Craig W. Miller has observed, the "concept of the atom" nonetheless "may best illustrate what Coleridge means" in his *Theory of Life* (1964, 78). It

---

[45] See Gower 1973. Specifically, *Naturphilosoph* Schelling's criticism of the determinist "atom of the mechanist philosopher" (Schelling 1858, 244; cf. 97–112) apparently played a major role in Coleridge's increasing weariness with the term in the late 1810s and early 1820s (Levere 1977, 360).

was Coleridge's way of defending the minded particle as a brightly shining "prism," not a dull, passive pebble propelled by the boot of physics. In his book, Coleridge sketches a universe composed of building blocks (1848, 46) ranging from "elements or simple bodies" to vast complexes formed from them. The poet's view of nature is combinatory and based on an assemblage that can be traced back to a microscopic, basic unit.[46] Still, these elemental building blocks are not the simplistic atoms that Schelling had dismissed as passive, static things but active expressions of energy: They are the result of a fundamental, electromagnetic polarity.[47]

"Matter," as Coleridge refers to his basic unit, is the "product, or *tertium aluid*, of antagonist powers. . . . Remove these powers, and the conception of matter vanishes into space" (1848, 55).[48] The elemental basis of nature is thus a "thing" that itself is an oscillating state of opposing energies—an odd dialectical structure, drawing from *Naturphilosophie*.[49] Its synthesis might connect each particle to its respective higher order, but one that is nonetheless in a temporary state, a constant flux that is always open to rearrangement.[50] In this logic, the particles of the brain do not move on strings pulled by external stimuli but tremble from within as a result of their constituent polarity between mind and matter.

While Coleridge's theism expresses itself in the belief that, ultimately, these opposing forces could be comprehended as united in the same,

[46] "We study the complex in the simple; and only from the intuition of the lower can we safely proceed to the intellection of the higher degrees. The only danger lies in the leaping from low to high, with the neglect of the intervening gradations" (Coleridge 1848, 42).

[47] If one were to look for echoes of contemporary chemists in his theory, a striking absence would manifest itself in the work of the highly influential Daltonian atomist Jöns Jacob Berzelius (1779–1848). Around the same time as Coleridge was composing what was supposed to be his grand philosophical treatise on nature (1812–1818), Berzelius was combining electrochemical and atomic theory to come to the groundbreaking conclusion that "every atom of a substance must possess electric polarity; the electro-chemical phenomena which are manifested when the atoms combine depended on their polarities; and the unequal intensities of the polarities of the atoms is the cause of the difference between their affinities" (summarized in Muirhead 2013, 244).

[48] Writing to English Swedenborgian mystic and science enthusiast Charles Augustus Tulk (1786–1849) in 1817, Coleridge states that his intention with this philosophical work is to "consider matter as a Product—coagulum spiritûs, the pause, by interpenetration, of opposite energies . . . I hold no matter as *real* otherwise than as the copula of these energies" (1959, 775).

[49] See also Hagner 2000, 220–23.

[50] On this level at least, Colerdige seems ahead of the chemists of his day. While even Berzelius understood atoms as matter *plus* energy, Coleridge seems to anticipate something more along the lines of energetic charges like protons and electrons (and perhaps even neutrons) that make up atoms as we understand them today. Of course, given that protons and neutrons are made of "quarks"—which are both matter *and* energy—the dichotomy between the two notions ultimately falls flat (see also Strassler 2012).

52 WRITING THE BRAIN

all-encompassing godly system (Miller 1964, 81), his idealism is none-theless a surprisingly *physical* one. This physicality, though, is not mere vitalism.[51] Taking up a familiar metaphor, Coleridge argues that his notion of active nature is *not* like "the breeze that murmurs indistinguishably in the forest [and] becomes the element, the substratum, of melody in the Æolian harp."

To Coleridge, an "organized body" does not channel passive nature into active life but itself is "a PRODUCT and representant of the power which is here supposed to have supervened to it" (1848, 67). Again, the poet is denigrating the eighteenth-century understanding of the harp (brain) as *making* nature sing (think), and instead embracing a reading of the harp as *making audible* a nature always already singing (being minded). With this novel conceptualization of matter that strives either toward diffusion ("the whole") or concentration ("individuation"), even the forces working on it are hardly spiritual: They are "repulsion" and "attraction," as well as "contraction" and "expansion" (Coleridge 1848, 55, 57). To Coleridge, all processes of nature, including mindedness, are elements of dialectical *ideas*, copulae of opposing energies that constitute entities/units as temporary arrangements only. Each element in nature is thus characterized by "the essential dualism of matter and the two-fold tendency to individuation" (Miller 1964, 86).

Using Coleridge's description of an idea as an orb[52] (see Figure 7), we can thus begin to map how mind and material are connected in Coleridge's enmattered idealism. In a contemporary letter, Coleridge explicitly names the seeming opposition of "Mind—Matter" as a "Copula" (point G in Figure 7) of nature (1959, 690). One part of this polarity might form the north-south axis (mind as contraction), the other the east-west axis (matter as diffusion),

---

[51] On vitalism and Romanticism (esp. Coleridge), see Mitchell 2013, esp. 86–92; Muirhead 2013, 118–29; Packham 2012; and Miller 1964. Mitchell rightly observes that *Theory of Life* aims to "to reorient, or repolarize the debate" over "transcendent vitalism versus materialism" and seeks a seemingly counterintuitive synthesis that brings the two together while "correcting" some of their faulty presuppositions (2013, 89).

[52] To make his heady concept somewhat more palpable, Coleridge, too, employs the favorite simile of atomic chemists like Dalton: the figure of a planetary orb. The center of this globe is the "oneness" of the idea as a "midpoint producing itself on each side" from opposing forces (i.e., in his image, the north and south poles) while at the same time being the "equatorial point of the two counteracting forces" (1848, 54). Thesis and antithesis are to Coleridge not points but lines, thus also extending between opposing forces and crossing at a point in the center that constitutes an idea, a temporary synthesis. And each line in itself can become a synthesis for a different idea—the result being the "total intussusception . . . of all in each" in nature (1848, 58).

NATURE'S MIND AND MIND'S NATURE 53

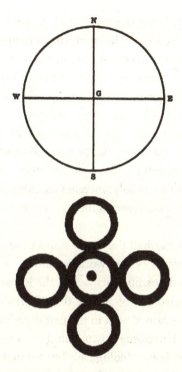

Figure 7 Coleridge's "orb" (in Miller 1964) versus Dalton's oxygen-hydrogen compound (Dalton 1808, 561).

and each in itself is again torn between solidity and particularization. For matter, these poles could be "magnetism" (contraction) and "cosmical electricity" (diffusion), Coleridge muses, with each side again constituted by polarity ("magnetism," in this case, constituted from the poles "carbon" as contraction and "azote" as diffusion) (1848, 57). Antithesis and thesis *mutually constitute* the synthesis through their polar opposition. A mind-boggling proposition, indeed.

In Coleridge's *Theory of Life*, then, nature is a temporary expression of the productive opposition of mind and matter, and neither pole can logically exist without its counterforce. Being everything, atoms are harping their tune as a duet between matter and mind. Wordsworth sensing mind radiating from rock, for instance, is thus not an act of playful transgression or momentary Romanticist synesthesia but a realization that the rock cannot be passive, cannot be a mere surface for poetic projection, but is instead an

54 WRITING THE BRAIN

active entity, constantly being produced anew as a synthesis between mind and matter. It is not *categorically* different from a brain, but merely different in the extent of the polarity that constitutes it. Fixity and flow—the opposing forces that Wordsworth explicitly senses "'mid the hollow depths of naked crags"—are indeed not contradictory but a productive, atomic polarity at work in brains and rocks alike. Coleridge even goes so far as to claim in different manuscripts that "all act is life" (in Muirhead 2013, 128) and that every thing encountered—be it a craggy hill or the west wind—has to be engaged as a *force* that is both object ("passive and material") and subject ("intelligence") at the same time, especially since, as Coleridge claims, "during the act of knowledge itself, the objective and the subjective are . . . instantly united" (1884, 885).[53]

Coleridge's theses never had the direct impact the budding philosopher had hoped for, with much of this work remaining in an unfinished state and only being released posthumously. Still, the Romantic notion of an elemental, a-priori-mindedness of nature that Coleridge tried to systematize was a force to be reckoned with in the first decades of the 1800s. To the scientists increasingly narrowing down "mind" to a physical process either temporarily arrested in or even biologically limited to the lobes in one's skull, Wordsworth, Shelley, Coleridge, and others seemed to reply that, certainly, mind was an *aspect* of matter—but in a different way than what "fatalists" might presume. Instead of dead matter reacting to outside impulses like wind moving passive harp strings, matter to these thinkers was everywhere always already mind, always already subject. And electro-dualist, atomic chemistry provided them with ample ammunition to combat the reductive drive of determinist anatomism and its physiological understanding of the brain that seemed to fetishize seventeenth-century mechanist psychology.

Still, forsaking the latter for mind-atoms and polar matter did not mean giving up on the favorite physio-poetical image of Hartley, Priestley, and others. The chime of the wind harp was too enticing, too suggestive to be abandoned so easily. Certainly, Aeolus's instrument had to be reframed, corrected, as Shelley was attempting. But some went even further. What was

---

[53] It is striking, then, how this Romantic's enmattered idealism—his eloquent (though laborious) charge against the "old" materialism of "dead stuff"—seems to come rather surprisingly close to New Materialism's embrace of actants and vital matter. This is an overlooked lineage, it seems, with many (often explicitly Spinozist) New Materialist works either sidestepping the Romantics altogether or at the very least ignoring Coleridge's philosophical theses. On New Materialism, see Coole and Frost 2010; Bennett 2010; and Latour 2005.

implicitly a chemist's harp in "Ode to the West Wind" (i.e., a tool that reveals characteristics of matter) became explicit in the many do-it-yourself chemical guides written for a public increasingly fascinated by the science. Akin to the popular, mass-produced wind harp, a Romantically inclined citizen could now produce "musical sounds . . . by the combustion of hydrogen gas." To do so, an often reprinted[54] experiment instructs its readers to

> bring [the glass tube depicted in Figure 8] down a few inches over the flame of the *philosophical taper* [glass filled with hydrogen gas; see *a* in the figure], and very strange but pleasing sounds . . . resembling those of an Æolian harp, will be immediately produced. By raising and depressing the tube, or by using tubes of different sizes, the intensity of the musical chord may be greatly varied. (Griffin 1849, 238)

With mind becoming an atomic property, the harp followed suit. Now, there was no longer any pretending that the wind itself was mere space traversed by mechanical physics; what is singing here is *hydrogen* (cf. Figure 7)—the lightest element on the periodic table that teases the harp to produce its art. Without atoms, *any* harp—cranial, wind, or chemical—would remain mute.

Thus, when Ralph Waldo Emerson, later in the century, praised this central metaphor for the Romantic mind, he made sure to include the lessons learned. "My friend," he advised the students at Tufts College in 1861, "stretch a few threads over a common Æolian harp, and put it in your window, and listen to what it says of times and the heart of Nature. I do not think that you will believe that the miracle of Nature is less, the *chemical power* worn out" (2001, 251). This careful reader of Coleridge's philosophy, who once famously labeled the "mind [a] part of the nature of things" (Emerson 1849, 60), truly knew that "these syllables that Nature spoke" via "the wind-harp" (2012a, 226) were neither ghostly whispers from the afterlife nor mechanical echoes of inert matter reluctantly vibrated into action, but literally "that which is held together" (Harper

---

[54] For instance, in Griffin 1860, 202, and—as variations on music produced by hydrogen combustion—in Dunant 1875, 294, or the anonymously published "Wind Instruments" 1881, 121. The "loud plaintive swell like that of the aeoilan harp" of this hydrogen experiment is also described in some schoolbooks, like that by H. L. Smith (1852, 178). Even the *Journal of Gas Lighting, Water Supply and Sanitary Improvement* would comment on the reaction's aeolian qualities ("On Some New Contribution" 1869).

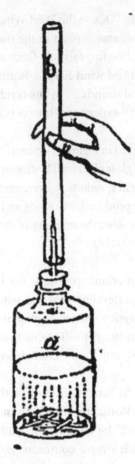

Figure 8 Griffin's atomic harp (1849, 238).

n.d., s.v. "syllable"): bonds of atomic matter, enabling song and thought in nature inner and outer.

***

Alan Richardson, in his monumental *British Romanticism and the Science of the Mind*, has proposed the term "Neural Romanticism" to account for the influence of the oft-neglected brain on Romantic thought. There must have been, he suggests, a "Romantic brain" (2001, 1) that one can unearth not only in the science of the early 1800s but also in its literatures. Still, as his sharp, analytical readings proceed, it becomes clearer and clearer that "Romantic"

NATURE'S MIND AND MIND'S NATURE   57

comes to function primarily as a temporal indicator (i.e., "Romantic period")[55] and that he locates these writers in a largely passive position, often merely reacting to their scientific contemporaries, including decidedly "unromantic" scientists of the preceding century. To Richardson, it was, for instance, the psychology of Erasmus Darwin (1731–1802) that taught Coleridge about embodied cognition, and Wordsworth about the imagination. In the same vein, Shelley Trower describes Coleridge's absorption of Hartley (2009). Still, to what extent are these notions really *Romantic*?

This issue is only exacerbated by both scholars noting a conscious move away from some of the very concepts now defined as cornerstones of "Romantic" notions of the mind. Is one to understand Coleridge's break from Hartley around 1804 (cf. Richardson 2001, 8–12), for instance, as a break from a "Romantic" concept (making him "less Romantic" all of a sudden)? Or is it perhaps a move away from science as such and into religious dualism (as Trower suggests)? What, really, did Romantic poets have to *add* to the understanding of the brain and cognition? Certainly, the science of the day was diverse and so were many of its Romantic readers, and one need read far to find disagreements over particular theories even between (and within) Wordsworth and Coleridge.

If one is to look for a discourse that can properly be labeled "Romantic" and that performs a sustained arch from the early fascination of many Romantic poets with mechanistic neuro-philosophy to high Romantic, idealist organicism, it is the debate over the Aeolian harp. Perhaps no other metaphor of the day so nicely helped frame, digest, and renegotiate the enmattered nature of the human mind (so persuasively argued across the scientific spectrum). A mechanical tool for entertainment at first, its melodious tunes initially came to sing to a young Coleridge (and a young Goethe) of the beauties of a passive mind. It was picked up by Sensationalist physiologists and became the perfect poetic mirror for the brain at the turn of the nineteenth century. Poets suddenly listened to poet-machines (at times built by poet-technicians) that mimicked and arguably surpassed their own creative minds. Their songs to the harp in turn infused the popular understanding of the brain in their day, providing a structuring metaphor that guided the anatomist's hands as well as the philosopher's pen.

---

[55] Hence, perhaps, Richardson's inclusion of Jane Austen (2001, 93–113), without seemingly needing even a brief explanation—generically or stylistically—of why this novelist so matter-of-factly seems to fit alongside Keats, Wordsworth, and Coleridge (the authors who populate the remainder of Richardson's work).

58   WRITING THE BRAIN

Yet the wooden object lining so many of these poets' windowsills also embodied the epistemological shortcomings of mechanistic philosophy. No creative mind could long sustain seeing itself perfectly echoed by a little box adorned with strings, a mere passive object. Instead of abandoning the wind harp, though, its poetic and scientific disciples dug deeper. Robert Bloomfield had already reminded his fellow Aeolianists that his products do not merely *produce* sound but, like a prism, disclose the constituent sounds of nature itself. Couldn't the same be true for the brain?

Given the extensive linkage between harp and brain that so permeated literary production in these years, it is no wonder that when the Romantics abandoned mechanistic physics for electrochemical atomism they brought the metaphor along. They debated it and it soon became their microscope, revealing nature itself—its constituent matter—as a minded entity, a living, active mass. The brain, thus, was not an exception to nature, but merely a "single mode of agency" through which nature's mind found expression. Every rock, every breeze seemed to oscillate with atoms of thought, particles of mind. One wonders whether Dalton ever read "Walking" Stewart or whether the latter read the former—but regardless, their convergent theories certainly grew out of the same soil, nourished by writers of science, philosophy, and poetry alike.

It is telling, then, that this deep connection between harp and brain still seems to survive today, albeit in a rather odd, return-of-the-repressed sort of way. Although we all still have a "psalterium" or "Lyra Davidis" nonchalantly lodged in our skulls, the brain proxy that is the Aeolian harp has largely receded from philosophy, literature, and science and has descended deep into the uncanny. If one were to dig through the twisted caverns of the collective unconscious of the internet, hunting for wind harps, one would soon discover "HAARP"—a US military facility shut down in 2014 and believed by conspiracy theorists to have been engaged in mind control via so-called "chemtrails."[56] To proponents of this theory, the condensation of airplane traffic in the air above, brings to them messages from a quite different, chemical wind harp: a much more sinister HAARP, manipulating the citizenry's brain chemistry through atomic infiltrators scattered by the winds onto their "indolent and passive" mind—the "dead thoughts" of Wordsworth's breezy "lyre" now a deeply gothic, paranoid distortion of an instinctual terror over the material limits of cognition. If one listens attentively, one might even hear the rattling of the Specter Harp's ghastly bones.

---

[56] For a brief summary of the "High Frequency Active Auroral Research Program" and the surrounding conspiracy theories, see Pappas 2014.

2

Split Brains, Doubled Minds

The Gothic's Bicameral Vision

Dreaming constitutes a pure state of unmediated and unthinking consciousness. . . . The somnipath [therefore] at once struggles against the chimera that is his own self and thus cannot be overcome.

—Friedrich Kittler, c. 1970[1]

I felt a Cleaving in my Mind
As if my brain had split -
I tried to match it - seam by seam,
But Could not make them fit

—Emily Dickinson, c. 1864

The Sleepers

In the late 1760s, perhaps the early 1770s, a strange occurrence transpired on the family farm of militia captain Joseph Miller (1727–1803) of West Springfield, Massachusetts. One of his four sons[2] began experiencing "fits," which caused a marvelous transformation. The first "paroxysm" transpired while the young man was bathing. After suddenly slipping into unconsciousness, he quickly arose a changed person, before waking and reverting to his old self shortly afterward. This strange pattern of behavior continued for around two years:

---

[1] Kittler 2015, 154; translation mine.

[2] Joseph Miller, named after his grandfather, had four sons, all born in the 1750s: Aaron J., Leonard, Joseph, and George. It is unclear which son was affected, but Aaron later became a respected physician (Williams 1908, 4–8).

Writing the Brain. Stefan Schöberlein, Oxford University Press. © Oxford University Press 2023.
DOI: 10.1093/oso/9780197693681.003.0003

60   WRITING THE BRAIN

> When a fit seized him, he would at first fall down; but in a moment or two rise, possessed of an agility far superior to what was natural. In two or three hours and sometimes sooner, the fit would pass off and leave him in his usual state, and, to appearance, in health. But what was most remarkable in his case, was the state of his mind. While he was in a fit, he perfectly remembered things which had occurred in all preceding fits; but nothing which happened in the intervals. . . . In the intervals, all his fits and everything which had passed in them were totally obliterated. . . . When one was present, the other was lost. . . . In short, he seemed to have two distinct minds, which acted in turns, independently of each other. (In Stiles 1933, 110)

Miller informed the town's priest, Joseph Lathrop, a recent Yale graduate, of these happenings. Lathrop would then convey an account of these strange "possessions" the president of his alma mater (and his former tutor), Ezra Stiles. Eventually, the priest's letter was printed in a number of periodicals, such as *The New-York Magazine* and *The American Museum*, and a few years later a familiar figure—none other than Benjamin Rush—would read the above passage from it to his attentive medical students.

What was what happened to the captain's son indeed a case of a single person having two completely distinct, unrelated minds? Rush was skeptical, though he had heard of the theories of the Austrian anatomist Franz Joseph Gall (uttered before he would become infamous for his craniology) that the mind may be "a double organ, occupying the two opposite hemispheres" (1981, 672). Based on his work with "madmen," Rush informed his students that while the two states of mind exhibited by the militia captain's son might not be entirely separate, they constituted two parts of a mind that can never directly interact—one part is diseased and "depend[s] upon a higher degree of excitement" to come to the forefront (1981, 672). To Rush, thoughts and memories were physical *impressions* upon the brain—but some of these may be so deep that "preternatural excitement" is required to access them (1981, 672). There can thus exist two separate sets of memories, thoughts, and beliefs in a single, diseased brain—located in different regions (perhaps: different hemispheres), with one set active only when its counterpart is inactive.

What may have made this assessment troubling to Rush's audience was another statement from the famous physician: "Sleep is a disease" (1981, 674). Still, it is a disease with a purpose:

SPLIT BRAINS, DOUBLED MINDS   61

In persons who labor during the day, [sleep] abstracts excitement from the brain, and diffuses it through the limbs; and in persons who pass the day in study, it abstracts excitement from the limbs and diffuses it through the brain. (1981, 675)

This "form of fever" (as Rush terms sleep) serves to balance out the distribution of "excitement" (heat, blood flow) over the course of the day. In Rush's heroic medicine, each part of the body "disproportionally exercised" during the day is granted replenishment from an opposing system of the body in sleep (1981, 675). A disease acting in kind with sleep may then overexcite a certain part of the mind and bring about a separate, insane version of one's self with its own memories and reasoning, located in a different part of the brain (1981, 672). There may thus be a segment of one's psyche that the waking self is unaware of and that may take over one's body at night.

Listening to these unsettling theories was a young medical student named Elihu Hubbard Smith (1973, 7). A "very close friend" of Smith's (Axelrod 1983, 21) may have been attending by proxy: Charles Brockden Brown. Brown showed great interest in the themes of these lectures[3] and only a few years later announced, but never fully published, a now-lost novella on the topic of somnambulism (1798's *Sky-Walk; or, The Man Unknown to Himself*), for which Smith, who often assisted Brown (Krause 1998, 908), supplied the necessary medical information (Barnard and Shapiro 2006, xviii). Brown published yet another novel on a Rushean topic—the yellow fever epidemic of 1793[4]—in the same year as *Sky-Walk*. Only a few months later, Brown returned to the theme of sleepwalking in what remains the most successful piece of contemporaneous American fiction on the issue: *Edgar Huntly, or, Memoirs of a Sleepwalker* (1799/1800).

That the question of the somnambular duality of the human brain was on Brown's mind becomes clear through a series of "Queries" he published in a periodical in June 1798, soliciting advice from his readers on several subjects that would become major themes in his various novels appearing in

---

[3] Brown's biographer agrees: "If [Brown] did not know Rush's work directly, then certainly he must have heard of it" (Axelrod 1983, 21).

[4] *Arthur Mervyn; or, Memoirs of the Year 1793* came out a year prior to *Edgar Huntly* and a year after Hubbard Smith had died of yellow fever in New York. A 1793 diary entry by Smith might suggest that Brown was aware of Rush at the time and communicated with Smith about him: "Recd. a letter from C. B. Brown. The accounts of the state of the sickness in Phila. appear to be very contradictory. Dr. Rush writes to his son that the disease is on the decline" (Smith 1973, 352). Whether the information about Rush was contained in Brown's letter (which is not extant) or was a separate piece of information remains unclear.

62   WRITING THE BRAIN

the months following. Queries 7 to 10 deal explicitly with mind-materialism and sleep:

7. Does impenetrability belong to matter?
8. Can the mind contemplate two ideas at the same time?
9. Is sleep the suspension of thought or only of memory?
10. Is sleep a perfection or a defect in the human frame?—If the latter, is it curable? (Brown 1798, 233)

Brown was obviously familiar with Pennsylvania's famous new resident, English mind-materialist Joseph Priestley. Priestley, one of the many co-discoverers of oxygen, had notoriously argued that "impenetrability" was *not* a defining feature of matter, and that the prevailing theory that impenetrable matter cannot produce mind in the same locus was thus void.[5] From the question of whether a single substance can exist in different states simultaneously, Brown proceeds to consider *ideas*: Could a brain hold two different trains of thought at the same time (Query 8) or would it have to enter a different state of consciousness for that—as in sleep (Query 9)? This makes Brown wonder, clearly echoing Rush's description of sleep as a beneficial fever: If sleep is a negative that results in a positive, would sleep's *thoughts* be considered good or bad (Query 10)? Whatever replies Brown may have received, by the time he composed *Edgar Huntly* the answers were in place: Yes, in some cases the mind *can* produce two distinct trains of thought, but one has to be carried out during sleep—and, like in Rush's theory of slumber as a necessary evil, these nightly thoughts are violent and unsettling but ultimately necessary to bring about closure.

Brown and Rush here participate in what Hagner calls a "*naturphilosophical* milieu" in which "parameters like the difference between the front and back of the brain, its doubled nature, and its asymmetry, are hooked up to established psychological and anthropological categories like reason and emotionality, primitiveness and the capacity for civilization, masculinity and femininity, or conscious and unconscious life" (2000, 213, translation mine). Keeping in line with the theme of duality, *Edgar Huntly* features not one but

---

[5] Focusing his chemical research on gases (i.e., matter that clearly permeates), Priestly not only argued against "impenetrability" as a defining feature of matter, which in turn allowed for matter to have *properties* or *qualities* (like "mind"), but also claimed that, since man's thinking is made of particles of thought, associated into ideas, an "immaterial" holistic substance ("Soul"), as proposed by dualists, couldn't possibly produce mind—only complex matter could. On Priestley's objections to impenetrability, see Priestley 1782, 16–28; Dybikowski 2008, esp. 100–103; Shook 2012, 848.

two characters haunted by their own somnambular doubling. The epistolary tale sees both its eponymous main character and the Scots-Irish servant Clithero take to sleepwalking to come to terms with harsh frontier life, evil twins, disputed rights to Native land, and the violent demise of Huntly's friend.[6] The story begins with the narrator encountering a shadowy figure digging by the very elm under which said friend had been slain. The oddness of the pale figure's behavior and his apparent inability to see Edgar leads him to conclude:

> It could not fail to terminate in one conjecture, that this person was *asleep*. Such instances were not unknown to me, through the medium of conversation and books. Never, indeed, had it fallen under my own observation till now, and now it was conspicuous, and environed with all that could give edge to suspicion and vigour to inquiry. (Brown 2006, 10)

Long before discovering how his own perturbed self led him to sleepwalk, Edgar now sets out to follow this shadowy figure (later revealed to be Clithero)—but ultimately loses him, when the sleeping man eludes him by entering a subterranean system of caves.

Much to his shock, Huntly will later find himself lost in these very caves as a result of his guilt-driven somnambulism. Waking suddenly in unfamiliar surroundings, he narrates:

> My return to sensation and to consciousness took place in no . . . tranquil scene. I emerged from oblivion[7] by degrees so slow and so faint, that their succession cannot be marked. When enabled at length to attend to the information which my senses afforded, I was conscious for a time of nothing but existence. It was unaccompanied with lassitude or pain, but I felt disinclined to stretch my limbs or raise my eyelids. My thoughts were wildering and mazy, and, though consciousness was present, it was disconnected with the locomotive or voluntary power. (Brown 2006, 107)

---

[6] Waldegrave, likely inspired by Brown's friend Smith (Crain 2001, 136).

[7] Brown's split between what could be termed "raw cognition" (sensation) and the sense of self (consciousness) is striking: Huntly's thinking apparatus awakes *before* his self does, again underscoring its physiological autonomy from Huntly's reading of himself as himself. To borrow from Garrett Stewart's reading of Rogue Riderhood in *Our Mutual Friend*, in this brief moment the cognition in Huntly's head is "raw biology not yet subsumed to his familiar brutish psychology, [and] the character is held in 'abeyance'" (Stewart 2015, 128).

64 WRITING THE BRAIN

At a later point of the story, and after having realized that he was a victim not of kidnapping but of his own brain, Huntly concludes:

> Thus it was with thy friend. Hurried on by phantoms too indistinct to be now recalled, I wandered from my chamber to the desert. I plunged into some unvisited cavern, and easily proceeded till I reached the edge of a pit. There my step was deceived, and I tumbled headlong from the precipice. The fall bereaved me of sense, and I continued breathless and motionless during the remainder of the night and the ensuing day. How little cognizance have men over the actions and motives of each other! How total is our blindness with regard to our own performances! Who would have sought me in the bowels of this mountain? (Brown 2006, 185)

The radicalness of this moment is easily lost to readers in a post-Freudian world but would have been startling to Brown's contemporaries: Edgar Huntly is describing the brain as a *necessary*, but not *sufficient*, condition for "consciousness" (a term repeated thrice in a mere five short sentences). Mind, in all its complexities, can exist with or without consciousness, rendering non-/unconsciousness a much more active state than was generally believed.

The affliction of the two sleepwalkers is not a mere fancy of the psyche, though, but a material event. Upon realizing what had befallen him, one of the men, Clithero, even experiences actual spasms of the brain, when trying to comprehend (and narrate) its actions:

> His complexion varied from one degree of paleness to another. His brain appeared to suffer some severe constriction. He desired to be excused, for a few minutes, from proceeding. In a short time he was relieved from this paroxysm, and resumed his tale with an accent tremulous at first, but acquiring stability and force as he went on. (Brown 2006, 46)

Echoing the heroic medicine of Rush, the brain of Clithero contracts like a blood vessel, sending him into a short "paroxysm" that brings about—albeit for only a second—the pale visage[8] of his other, somnambular self.

A century before Freud would propose the existence of an "unconscious," Rushean notions of the brain enabled Brown to describe a mind that does

---

[8] "Persons who are affected with this disease," Rush later writes of sleepwalkers, "appear pale, and covered with profuse sweats" (1812, 303).

not require consciousness: Brown's sleepwalkers are able to navigate complex paths, react to visual cues, lock and unlock doors, recall their past actions (Clithero, for instance, keeps returning to the elm), and perform complex, multistep actions like digging and then refilling a hole. As a sleepwalker, Huntly has "sense," "locomotive [and] voluntary power"—things that allow him to speak of the actions performed as a result of these powers as *his* ("I wandered," "I plunged," "I tumbled," etc.) even though they happen without any awareness of or contribution by his "self." This leads him to soliloquize that one's "blindness with regard to our own performances" equals the "cognizance . . . men [have] over the actions and motives of each other." The mind, to Brown, is a stranger to itself, and the "I" generates motives post hoc by interpreting its actions. Lost in the caves of their skulls, Brown's characters can merely speculate over the state of their minds: Where Wordsworth experiences awe and enmattered interconnectivity, Brown finds "consciousness" to be a mere visitor, perhaps even an unwelcome one. Identities are thus mere "phantoms" a brain "makes appear" (from *phainein*, "to bring to light, make appear").

Still, an aura of danger lingers around the nighttime personas of these sleepwalkers. Sleep, after all, is a state of "disease." Upon first spotting the somnambule digging below the elm, Huntly makes a case for what sounds so much like "repression" that it has rendered this story a particular favorite among psychoanalysts:[9]

> The incapacity of sound sleep denotes a mind sorely wounded. It is thus that atrocious criminals denote the possession of some dreadful secret. The thoughts, which considerations of safety enable them to suppress or disguise during wakefulness, operate without impediment, and exhibit their genuine effects, when the notices of sense are partly excluded and they are shut out from a knowledge of their entire condition. (Brown 2006, 17)

And, indeed, while Clithero has not actually committed a murder, his sleepwalking delusions almost drive him to stab the love of his life in a case of mistaken identity.

Upon realizing what he was about to do—and being caught redhanded—Clithero attempts the only escape possible to him: "Murder was

---

[9] Representative pieces through the decades include Hustis 2003; Fiedler 1997; Smith-Rosenberg 1993; Hedges 1974.

66 WRITING THE BRAIN

succeeded," in his mind, "in an instant, by the more detestable enormity of suicide" (Brown 2006, 60). Suicide, too, flickers across Huntly's brain when realizing the helpless situation in which he finds himself at one point in the story: at the bottom of a dark pit, at the mercy of whoever put him there, unable to get out by himself. Still, considering the dexterity of Brown's sleepwalkers, one has to wonder whether Huntly, this consistently unreliable narrator, is indeed telling the truth. Was his "step . . . deceived," so he accidentally fell (as he claims later), or did he, too, attempt suicide in what he later realized was a sleepwalking state? Certainly, Huntly's subsequent transition from waking in an apparent kidnapping situation to considering ending his life transpires rather quickly, perhaps instead hinting at a *lingering* desire to end his own life. Was Huntly's mind, too, "sorely wounded"? If so, he is careful to not let his audience know the exact nature and extent of his melancholia.[10]

For a contemporaneous audience, at least, suicide would have been at the forefront of their minds when reading about sleepwalking. Indeed, in the early nineteenth century, in the United Kingdom as well as the United States, "sleepwalking" became a common trope in order to avoid discussing this stigmatized subject outright[11] (cf. Bell 2012). This, in turn, reinforced fears that one's nocturnal self might just decide to jump out of an attic window when the (primary) consciousness had surrendered control. While notices about sleepwalking *murderers* would remain scarce until midcentury,[12] major newspapers in the early decades of the 1800s would frequently run short notices like these:

> Mr. Martin, of Sutton, near Holbeach, being on a visit at Boston, threw himself on Sunday night, whilst walking in his sleep, out of his bed-room window into the street, a height of 20 feet. ("Cambridge, April 4th" 1811)

---

[10] Besides the sadness over his friend Waldegrave's passing, Huntly mentions no major event or act that may be haunting his mind. Unlike Clithero, Huntly never believes himself to be a murderer, and his nighttime rambles do not push him to attempt to kill anyone—except, perhaps, himself.

[11] Another reason why family members, especially, would favor a sleepwalking assessment is perhaps best summarized by this debate from the second half of the 1800s: "The question has been raised whether falls from a height in sleep-walking can be considered suicide in the meaning of life insurance policies. But it has been ruled that the proviso against suicide only includes intentional killing" (Woodman and Tidy 1877, 857).

[12] This had somewhat changed by 1846, after the murder of a sex worker, Maria Bickford, by Albert Tirrell, who claimed to have killed her while sleepwalking—and was acquitted. Stories like "The Sleep-Walker; or, Lady Theresa's Trials" (published in the *London Journal* and the *Australian Journal*) directly respond to that case. Tirrell later actually approached Edgar Allan Poe in the hopes of founding a journal with him, but Poe declined.

[Elvira Ayres] retired to rest last Thursday evening, well and cheerful, when Mr. Bennet arose, on Friday morning, he found her missing. He searched the house, and, not finding her, grew alarmed—he searched the barn and the well, then raised the lid of the cistern, where he found her and drew her out a lifeless corpse. . . . It was learned from her mother that she had been known to get up in her sleep . . . [and] had, in several instances, awaked and found herself several rods form the house. ("Dangers of Sleepwalking" 1824)

Wednesday morning, about three o'clock, a block cutter, of the name of Buchanan, who resided in Tradestown, Glasgow, was found lying on the street shockingly bruised. He was taken to the Royal Infirmary, where he died about noon. . . . [I]t is supposed he had fallen out of the window, in a fit of sleep-walking. ("Magistrates of Irvine" 1816)

Accident.—Mr. John A Knapp, late of Cincinnati, came to this death on Thursday night last, by rising in his sleep, and walking out of the window of a garret room, in which he usually lodged. ("Accident" 1825)

The idea of "a person apparently well [going] to bed without manifesting the slightest tendency to self-destruction" (as a midcentury psychiatric journal summarized it) and killing themselves before waking up (Winslow 1856, xxi) was certainly a troubling prospect to anyone concerned at all about the intentions of their nightly selves. While medical ideas on why sleepwalking occurred generally echoed Rush, the *experience* of having (being?) a suicidal nighttime doppelgänger generally remained a mystery, with even Brown resorting to having his characters (like Clithero) speculate on their motives after the fact.

Dialoging the Self

One of the few literary attempts at grasping the *conscious* experience of having two contradictory selves may be Alfred Tennyson's poem "The Two Voices," an internal dialogue on suicide, written in the 1830s. It was likely influenced by a period of depression following the death of a close male associate. The iambic conversation, which notoriously fails to come to a synthesis

68 WRITING THE BRAIN

of the speaker's selves, reads so cranial[13] that mind-materialist Herbert Spencer (1820–1903), two decades after its writing, sent the poet his *The Principles of Psychology*, arguing that "The Two Voices" echoed his monist psychology, and hoping (in vain) for the then-famous writer's endorsement (Tate 2009, 61).

The dialogue between the speaker and a "silent voice" (both rendered in direct speech) seems to mimic the meditative, semi-somnambular practice of Tennyson that he described later in life. Ever since he was a little boy, the poet claims, he had the habit of self-inducing a "kind of waking trance" that caused him to experience how "out of the intensity of the consciousness of individuality, the individuality seemed to dissolve and fade away ... [and] the loss of personality (if so it were) seem[ed] no extinction but the only true life" (in Almond 1982, 27). This theme forms the core of the conflict in "The Two Voices": Is there hope for transcendence in a mind constituted by matter? Is the death of self "merely" an end or instead a recycling of mind in a grander, universal scheme? Tennyson's primary voice argues for the wonders of life, but the other, "silent voice" keeps undercutting his arguments and proposing death as the better, grander alternative.

The narrative situation of this internal debate (taking place, as the speaker puts it, "in my mind" [Tennyson 1994, 173]) is never made explicit but hints at a dreamlike scenario: the dialogue takes place at "night," the suicidal voice mocks the speaker's inability to continue sleeping (Tennyson 1994, 173), whereupon the speaker urges *it* to return to sleep ("Go, vexed Spirit, sleep in trust" [Tennyson 1994, 175]). The "silent voice" even repeatedly calls its other half a "dreamer." The poem thus seems to narrate the "wildering and mazy" state (Brown again) where the two selves may meet: the exact moment of drifting in or out of sleep, where Tennyson's speaker can arrest his nocturnal, suicidal self and interrogate it. At a later point in the poem, this situation is linked rather directly to the medical concept of sleepwalking:

> As here we find in trances, men
> Forget the dream that happens then,
> Until they fall in trance again.[14]

---

[13] "The Two Voices" even echoes the Romantic trope of brain-as-harp: "'I may not speak of what I know.' / Like an Æolian harp that wakes / No certain air, but overtakes / Far thought with music that it makes" (Tennyson 1994, 184).

[14] Again, we find the notion of two neurologically distinct sets of memories contained in one brain—one produced during sleep(walking), one during waking. A similar case can be found

SPLIT BRAINS, DOUBLED MINDS   69

> So might we, if our state were such
> As one before, remember much,
> For those two likes might meet and touch. (Tennyson 1994, 182)

Two persistent selves, one active when the other is inactive, with their own sets of memories, meeting in the liminal space between sleep and waking—a "waking trance,"[15] as Tennyson calls it—to commune in the hopes of consolidating their minds. "Sick art thou—a divided will," the voice mocks (Tennyson 1994, 175).

Still, abandoning one voice for the other, or having them agree, is impossible: The "other" voice, too, is of the speaker's mind. And, though drawing from different memories, the two voices are still located in the same brain and cannot be dismissed or washed away with Romanticist pathos. After echoing the joys of oneness with nature, the speaker thus concludes the poem by gazing at the one, true marvel: the divided mind.

> I marvell'd how the mind was brought
> To anchor by one gloomy thought;
> And wherefore rather I made choice
> To commune with that barren voice,
> Than him that said, "Rejoice! Rejoice!" (Tennyson 1994, 184–85)

Instead of chiming in with the chorus that calls for the speaker to "rejoice" in life, it is suicidal introspection that is the real source of wonder in the poem. Ultimately, it is the marvel of the brain that keeps the speaker going, pledging to return to his suicidal self again and again to experience the intensity of its death-bound convictions.

Tennyson has ventured into his "Gothic mode" here (Ackerman 2009, 85–135)—and one of the masters of the genre was paying close attention. Back

---

in Wilkie Collins's 1868 *The Moonstone*, which actively embraced the idea of two separate sets of memories as a plot device. Nicely enough, Collins informs us of his source: "The book in your hand is Doctor Elliotson's HUMAN PHYSIOLOGY; and the case which the doctor cites rests on the well-known authority of Mr. Combe" (Collins 1868, 190). The anecdote of the drunken Irish porter he is referring to can be found in Combe's *Essays on Phrenology* (1830, 521) and is, indeed, reprinted in John Elliotson's *Human Physiology* (1835, 646).

[15] "Waking trance" was often used as either a synonym for "somnambulism" (Mayo 1849) or as a descriptor for a particularly "intense" form of it (Monro 1851, 44). "Walking trance" was the term favored by mesmerists to describe the condition, especially at a time when what is today understood to be hypnosis and sleepwalking were conceptualized as like states.

70 WRITING THE BRAIN

in Philadelphia, a fervent admirer of Brown and Tennyson[16] was composing a short story on this very theme that would appear in August of 1844 in the *Columbian Magazine*, and about a year afterward, in the *American Phrenological Journal* and the *London Popular Record*. The *Record* retitled the piece "The Last Conversation of a Somnambule"—a title change to which its author, Edgar Allan Poe, objected since it gave away the end of the story (Mabbott 1978). Poe preferred its original title: "Mesmeric Revelation." Written around the same time as his article detailing his "reverence unbounded" for Tennyson (Poe 1850, 611)—and specifically for "The Two Voices" (Poe 1845)—it is Poe's ghoulish twist on "commun[ing] with that barren voice."

Of course, Poe (see Lind 1947), even more so than Tennyson (see Sait 1974), was a believer in mesmerism—and the short tale is also brimming with this occultist pseudoscience (considered so even in the 1840s) that combined a form of hypnosis with a vitalist belief in a pervasive life-force termed "animal magnetism." Still, while the underlying assumptions behind the prevailing medical views on sleepwalking and those held by "animal magnetists" differed significantly—one set drawing on a physiological monism, the other on an electromagnetic dualism—their descriptions of the state at the center of *Edgar Huntly* were so congruous that "waking trance" and "somnambulism" were popularly understood as a difference not in kind but in terminology. As Rush's biographers put it: "Mesmer claimed credit for discovering induced somnambulism" (Carlson, Wollock, and Noel 1981, 401).[17] Mesmeric hypnosis ("trance") was thus also often referred to as "artificial somnambulism" (St. Dominique 1874).

Such an "artificial somnambulism"—here termed "sleep-waking" (Poe 2006, 632)—is at the center of Poe's tale. Like the internal debate in Tennyson's poem, Poe's dialogue deals with the immortality of matter versus the death of the individual. The crux of this debate may be best illustrated by recalling the defense of death as the *release* of matter in Tennyson:

> This truth within thy mind rehearse,
> That in a boundless universe
> Is boundless better, boundless worse.

---

[16] See Carter 1953.

[17] "I reject the futile pretentions of Mr. Mesmer . . . he has absurdly called animal magnetism. But I am willing to derive . . . advantages from his deceptions" (Rush 1789, 249).

SPLIT BRAINS, DOUBLED MINDS   71

> Think you this mould of hopes and fears
> Could find no statelier than his peers
> In yonder hundred million spheres? (Tennyson 1994, 172–73)

Tennyson's waking self fails to ultimately embrace this transcendence in universal matter—but Poe's "colloquy," occurring between a sleep-waker" and a mesmerist (2006, 632), comes to a different conclusion. Not only does the somnambule philosophize quite overtly here, but he underscores his lecture with a Gothic punch: At some point in their conversation, the sleeping man had quietly surrendered to death and had been "addressing [the narrator] from out the region of the shadows" (Poe 2006, 639).

The theories at play in Poe's tale are a heavy helping of Coleridgean notions of minded matter and Rush's brand of medicine, with the language of mesmerism employed largely for Gothic ornament.[18] The medical perimeters are set by Rush: In keeping with his definition, the sleep-waker's "intellectual faculties are wonderfully exalted and invigorated" (Poe 2006, 632),[19] and he had previously been under a steady regimen of heroic treatments.[20] The inquiry then begins with a confirmation of the subject's state of consciousness:

P. Are you asleep?
V. Yes—no I would rather sleep more soundly.
P. Do you sleep now?
V. Yes. (Poe 2006, 634)

Freed from the "idiosyncrasy of [his perceptual] organs," the subject then contemplates the prospect of his death (he is suffering from tuberculosis) and the nature of mind. To the narrator's surprise, he does not seem perturbed. Instead of bemoaning his state of affairs, the man offers heady theorizations on spirit and matter—all based on the presupposition that the somnambule's mind is more readily able to reflect on itself since it does not find itself entrapped by the memories, impressions, and feelings of the waking self.

Poe, at the time, was enamored of Coleridge's works, leading some critics to observe that the British poet had become the "guiding genius of Poe's

---

[18] During the conversation, it becomes clear that the *narrator* is a "mesmerist"—but the sleepwalker is far from it, and persistently chastises his interviewer's narrow concepts and terminology.

[19] Rush speaks of "preternatural excitement."

[20] Especially the "application of mustard to the nervous centres" (Poe 2006, 632), a favorite of Rush's (1812, 102); see also Rush 1981, 225, 569.

72 WRITING THE BRAIN

entire intellectual life" (Stovall, in Schlutz 2008, 195). Even Coleridge's rather obscure *Hints Towards the Formation of a More Comprehensive Theory of Life* has been tied to Poe's work in the 1840s (Mills 2006, 9). Drawing heavily from his British master, Poe's somnambule thus proposes that "mind" is the principle behind all matter, making up the basic building block of the "atomic constitution" of the universe (its quarks, so to speak). Everything, this "silent voice" argues, is vibratory, atomic matter, constituted as mind, and associated into larger systems, particles, and bodies:

> There are gradations of matter of which man knows nothing; the grosser impelling the finer, the finer pervading the grosser. . . . These gradations of matter increase in rarity or fineness, until we arrive at a matter unparticled—without particles—indivisible—one and here the law of impulsion and permeation is modified. The ultimate, or unparticled matter, not only permeates all things but impels all things—and thus is all things within itself. This matter is God. What men attempt to embody in the word "thought," is this matter in motion.
>
> . . . Motion is the action of mind—not of thinking. The unparticled matter, or God, in quiescence, is (as nearly as we can conceive it) what men call mind. . . . [T]he unparticled matter, set in motion by a law, or quality, existing within itself, is thinking. (Poe 2006, 634–35)[21]

Harkening back to the problem of "permeability," the voice observes: "There is no immateriality—it is a mere word. That which is not matter, is not at all—unless qualities are things." The totality of the smallest, indivisible particles of matter may be called "God," its smaller, finite groupings "mind." When Tennyson's somnambular voice urges its speaker to "rehearse" in its mind the notion of a "boundless universe," it may just be hinting at this concept of the pervasive mindedness of all matter, hence rendering death a *return* to mind writ large, not its demise.

---

[21] The theories expressed here seem to mirror Poe's own. In a letter of July 1844, he writes: "There is no such thing as spirituality. God is material. All things are material; yet the matter of God has all the qualities which we attribute to the Spirit; thus the difference is scarcely more than words. . . . Man and other beings (inhabitants of the stars) are portions of this unparticled matter, individualized by being incorporated in the ordinary or particled matter. Thus they exist rudimentally. Death is the painful metamorphosis" (1966, 260).

These musings, of course, come with a narrative gut punch. In order to think through the rather lengthy, often convoluted, and expressly "philosophical" (the somnambule even defines terminology) theories put forward, a reader has to temporarily block out the narrative frame and instead focus intently on the line of reasoning of the speaker—rendering the revelation of his demise at the end a surprisingly visceral moment. Like Poe tracing the train of thought of dead Coleridge,[22] his own readers had been intimately engaged with the mind of a dead man, walking through its corridors, observing its convolutions, without realizing it to be a tomb. In a sense, though, this externalization of mind onto the physical matter of the page underscores Poe's/ Coleridge's point: If "thought" is minded "matter in motion," rehearsing this motion in one's head is truly akin to speaking to/as the dead. Reading, it seems, is always somnambular, always a dialogue of "two voices" in one mind.

But where, cerebrally speaking, would the somnambule's inner Coleridge be located? To the mesmerist in the story, this remains unresolved, except for the fact that this other cognition is removed from the "idiosyncrasy of [the perceptual] organs" (Poe 2006, 638). Rush had previously hinted that the second self inhabits some other region of the brain, but, since Rushean physiology theorized that all regions of the cerebrum could potentially produce mind,[23] he never speculated on its exact locus (though he was enticed by Gall's suggestion that it might lie in a different hemisphere). Poe, instead, had his own, peculiar theory. A few years before his mesmeric tales, he had likely penned a short treatise, *Philosophy of Animal Magnetism*, published under the guise of "a gentleman from Philadelphia."[24] Unsurprisingly, this

---

[22] The somnambule even shares a cause of death with Coleridge: heart failure in combination with a respiratory disease ("phthisis . . . with acute pain in the region of the heart"). Writing about the experience of reading Coleridge, Poe exclaimed: "In his . . . works we have beheld the mind . . . of the man" (1902, 51–52).

[23] "We sometimes find upon dissection the same parts, which have been found to be diseased or destroyed in mental diseases, to be equally diseased or destroyed in some people; and yet no mental disease to follow it. In these cases the deranged faculty, or even the whole mind, appears to be translated to an adjacent and sound part of the brain. . . . It is possible, in those cases in which the mind exists with nearly a total destruction of the brain . . . , the mind may be translated to the medulla oblongata, or to some of the larger nerves" (Rush 1981, 476).

[24] Major voices in Poe criticism of the twentieth century have argued against attributing the pamphlet to Poe. Laverty summarizes: "One book on mesmerism, *The Philosophy of Animal Magnetism*, by A Gentleman of Philadelphia, has been attributed to Poe by Joseph Jackson [in 1928]. But the book is probably not Poe's. The style is not his; some of the ideas are not his, for example, the praise of Colonel William L. Stone, for whom Poe had little friendship. It is the opinion of Dr. T. O. Mabbott, too, that 'with the writing of this book . . . Poe had nothing to do. . . .' This is part of a quotation written on the flyleaf of the copy in the Duke University Library" (Laverty 1951, 288). Subsequent scholarship has echoed Mabbott's view and dismissed Jackson's attribution for lacking "convincing evidence" (Hayes 2013, 322). A statistical authorship attribution of 2017 has backed up Jackson's proposition and added weight to the idea of Poe as the author of the pamphlet (Schöberlein 2017, 651–53).

74 WRITING THE BRAIN

eighty-four-page pamphlet is brimming with commentary on "somnambulism" (Poe 1837, 34–38, 41–43, 48–49, 54–57, 63–69) and gets so specific about the physiological underpinnings of one's inner, "silent voice" that one can comprehend the reluctance of Poe critics to embarrass the "Master of the Gothic" (Fisher 2002, 72) with its authorship.

Yes, in somnambulism consciousness awakes in a different cerebral region, Poe here asserts—but that region isn't even in the skull. Instead, we find the somnambule thinking with his or her solar plexus:[25]

> Now, it is a singular fact, that in many cases of catalepsy and somnambulism, the usual organs of the senses have been found to be *entirely dormant*, and the *seat of general sensibility transferred from the brain, to the region of this ganglion, or cerebrum abdominale*. Does not this circumstance suggest some distinction hitherto not sufficiently investigated, between the *intellect* and the *sensibility*—between the cerebral and the ganglionic systems of the nervous energy? (Poe 1837, 34–35)

Human cognitive duality, on a very material level, was to Poe a physiological fact,[26] not a mere philosophical suggestion. Each person contained within him- or herself a hidden "other" that thought different thoughts and experienced a vastly different reality—a reality that was at times superior to one's primary reality, led astray by the impression of the senses. If Edgar Huntly felt terror over being ousted from his own mind, Poe saw it as an entirely justified move: One's primary consciousness, he claimed, was a delusion, leading him to attack the overreliance of the rationalists on their senses with a vigor almost equal to that of "Walking" Stewart.

As outrageous as Poe's claims about a solar plexus–mind might sound, the fundamental thesis behind Poe's mesmeric verbiage echoes contemporaneous attempts by psychiatrists and anatomists to begin to locate where *exactly* one might find mind in brain. If we are all "double Dupin[s],"[27] where would our Faubourg Saint-Germain be? While sleepwalking cases have made an intriguing argument for the existence of another, dormant self, they provided little insight into the specific cerebral regions that might produce mind

---

[25] The solar plexus is still considered a significant region in contemporary occultist pseudoscience, a linkage likely going back to its role in mesmerism (Shermer 2002, 538).

[26] For an extended reading of Poe through modern neuroscience, with a particular focus on theories of mental duality, see Stamos 2017.

[27] "I often dwelt meditatively upon the old philosophy of the Bi-Part Soul, and amused myself with the fancy of a double Dupin—the creative and the resolvent" (Poe 2006, 376).

(except for underscoring that mind cannot be exclusively tied to a single region). To get to the root of the issue, a perhaps even more intriguing clinical population became the center of anatomical interest: people surviving with large portions of their brains missing.

Hemispheric Voices

No less than the head physician of Queen Victoria and her husband (as well as the personal physician to Princess Caroline) took it upon himself to hunt for the seat of consciousness. What Sir Henry Holland (1788–1873) formulated as a result was startling: The duality of the human mind—hinted at so strongly by somnambulism—may be reflective of the "brain as a double organ." That the human motor system was indeed cerebrally "dual" (i.e., the right side of the brain controlling the left part of the body and vice versa) had been known since the eighteenth century.[28] But now that mind had been more and more conclusively shown to reside in such an obviously dual structure as the brain, with its (at first glance) mirror-image chambers, barely connected by a few threads of tissue—why ought a scientist consider the mind a unitary whole? Studies of people functioning (albeit with slight motor issues) with merely half a brain seemed to suggest more brain may not be necessary to produce consciousness. So why are there two hemispheres? "It has been supposed indeed by some," Holland writes in his 1839 *Medical Notes and Reflections* in an apparent echo of Gall and Rush (among others), "that each side of the brain is separately capable of fulfilling [its] functions" (1839, 171). Explicitly referencing states like sleepwalking/sleep-waking, Holland agrees that they serve to demonstrate the possible existence of a state of "*double consciousness*," a concept he defines as

the mind pass[ing] by alteration from one state to another, each having the perception of external impressions and appropriate trains of thought, but not linked together by the ordinary gradations, or by mutual memory. (1839, 174)[29]

[28] Perhaps surprisingly, it was Swedenborg, prior to his turn to mysticism, who may have been the first person to formulate a modern theory of hemispheric lateralization—but never published it (Davis and Dean 2005, 121).

[29] Gregory Tate has rightly noted the "closeness of this definition to the 'trance' stanza of 'The Two Voices'" (2012, 48), though this book would suggest the poem does not merely "anticipat[e] later scientific theorizations of the double brain and mind" (2012, 49) but shares source material with them.

76 WRITING THE BRAIN

Was it possible to extrapolate fundamental insights about the workings of the hemispheric brain from such case studies of so "curiously contrasted states of mind" (Holland 1839, 174)? Holland was cautious:

> I am not sure that this subject in the relation of the two hemispheres of the brain has yet been followed into all the consequences which more or less directly result from it. Symmetry of arrangement on the two sides of the body is common indeed to all the organs of life. But the doubleness of the brain, like all besides pertaining to this great nervous centre, offers much more of curious speculation than the same constitution of other parts. That unity of consciousness in perception, volition, memory, thought, and passion, which characterises the mind in its healthy state . . . is singularly contrasted with the division into two equal portions of the material organ which more immediately ministers to these high functions. Yet, on the other hand, in the almost exact symmetry of form and composition of each hemisphere . . . we find argument not merely for the correspondence of functions, but even for that unity or individuality, of which consciousness is the interpreter to all. (1839, 161)

While the physician extraordinaire is ultimately skeptical of the idea of each person having two fully functioning minds in his or her skull, he is nonetheless willing to entertain the notion for a whole chapter, finally leading him to wax poetic about the longitudinal fissure: "Our existence may be said to lie on each side of this boundary; yet with a chasm between them so profound and obscure, that though perpetually traversing it in all the functions of life, we have no eye to penetrate its depth" (1839, 178). Like James Hutton gazing into the angular unconformity at Siccar Point, Holland senses a deep profundity in the gorge dividing each person's brain. Still, he ultimately decides to turn away from it, arguing that case studies of sleepwalking and of people missing a whole hemisphere constitute "evidence . . . insufficient in nature and amount" (1839, 171) to draw further conclusions.

Still, *Medical Notes and Reflections*—and especially this chapter—appears to have spoken to many people.[30] It prompted extended reviews in major periodicals like the *Guardian*, the *Eclectic Review*, and the *Athenaeum*, and

---

[30] Including Oliver Wendell Holmes Sr., who mentions the main thesis of the work in an installment of his *Autocrat at the Breakfast Table* (1891, 74).

SPLIT BRAINS, DOUBLED MINDS   77

was followed up rather quickly with a second British edition in 1840; as late as 1857, new editions appeared even in the United States.[31] The 1840s and 1850s were truly the decades of the dual mind, and Holland's book was met with open ears and eyes on both sides of the Atlantic. Holland wasn't alone in his ideas. In the early 1850s, for instance, physiognomist James W. Redfield gave several lectures in New York[32] on what newspapers described as theories of the "dual brain," stating Redfield "declares that the two lobes of the brain are male and female in their character" ("A Dual Brain" 1853). In 1850 Mauritian physician Charles-Édouard Brown-Séquard (who later relocated to England and the United States) began publishing his work on the duality of the nervous system, which led to his discovery of the spinal cord syndrome named for him, and ultimately resulted in his Hollandian *The Dual Character of the Brain* (1877). Other works of the 1840s loudly lambasted the all-too-common belief in the "pregnant absurdity of representing a man as consisting of *two egos*" (Moore 1846, 22) or defended the "Non-Duality of the Mind" to the scientific establishment (Cattell 1846). Holland and his fellow dualists, as careful as they were in wording, could not have summoned such ire all by themselves—for that, they needed the assistance of Arthur Ladbroke Wigan (1785–1847).[33]

In 1844, Wigan's *A New View of Insanity: The Duality of the Mind Proved* fully and proudly embraced what was only hinted at by Holland: Instead of a singular consciousness, Wigan proclaimed, there are two distinct minds contained in our skulls, with one often dominating the other. Wigan's book was hugely successful and widely discussed in the English-speaking world—from *Punch* and the *Athenaeum* to regional newspapers (such as the *Leicestershire Mercury* and the *Westmorland Gazette*). Its impact at the time can hardly be overestimated. "As amusing as a novel" (" 'Standard' Mortality" 1845), *A New View of Insanity* has been traced into the works of George Eliot (Shuttleworth 2001) and, quite obviously, Robert Louis Stevenson (Stiles 2006). Still, one of the major literary works on the duality of the human

[31] Holland's book was even translated into German, by Joseph Wallach, and published there in 1840.

[32] Redfield's office was located in New York City ("Physiognomy" 1850). In the publication that would result from his lectures, the two-volume *Outlines of a New System of Physiognomy* (1853), Redfield does not repeat this hemispheric claim—but does focus on cranial differences between men and women (especially in volume 2).

[33] On Wigan, see also Harrington 1987a, 23–29, and Clarke 1987. In a review of Clarke's work, Harrington writes about Wigan: "Today there is hardly a self-respecting split-brain or laterality research worker who would dare to admit ignorance of at least the name and essential thesis of this nineteenth-century medical man" (1987b, 245).

78 WRITING THE BRAIN

psyche, Emily Brontë's *Wuthering Heights*, may well be the first thorough literary engagement with these notions. Composed during the height of the Wigan boom of the late 1840s (1845–47), the novel's "intimate knowledge of the human psyche" radiating from almost every page has been noted by many scholars (Gorsky 1999, 173). From one of Brontë's first critics who saw the novel unfold as "shapes that come out upon the eye, and burn their colours into the brain" (Dobell, in Watson 1949, 246) to the plenitude of Freudian or Jungian readings that still shape the reception of this peculiar Gothic tale,[34] one of the key approaches to confronting the dualistic structures of meaning in *Wuthering Heights* revolves around an understanding of the human mind. With the seemingly overtly symbolic paring of the dark, scheming Heathcliff and the ephemeral, sickly Catherine as doppelgängers of sorts (cf. Moglen 1971, 391), Brontë's singular literary expression seems to invite, even force, such readings.

Theories of the dual mind not only surrounded Emily Brontë in the various periodicals her family subscribed to but may have entered the Brontë household in a number of other ways—most likely via John Forbes (1787–1861), a well-known London physician who published on the brain and with whom the family consulted on their daughters' health.[35] The *British and Foreign Medical Review or Quarterly Journal of Practical Medicine and Surgery*, edited solely by Forbes, had issued a lengthy review of Wigan's book in July 1845 ("[Review of a New View of Insanity]" 1845), criticizing the doctor's methodology while respecting the boldness of his theoretical claims—claims that Wigan's book specifically finds validated by the previous work of Henry Holland, who also happened to be the first cousin of writer Elizabeth Gaskell (1810–1865), a close friend and future biographer of Emily's elder sister, Charlotte. Dual brains all around.

Still, whether Emily (or one of her sisters) ever actually held Wigan's (or perhaps Holland's) publication in hand or not, there would have been no way of avoiding this decidedly Gothic and so easily popularizable view of the psyche. Each brain hemisphere, Wigan argues, "is capable of a distinct and separate volition, and . . . these are often opposing volitions" (1844, 26). Indeed, these halves have such discrete personalities that even their physical shape differs (1844, 29). Human personality to Wigan is the result of an

---

[34] E.g., Madden 1972; Dawson 1989; Newman 1990; Masse 2000; Tytler 2012.

[35] Charlotte Brontë wrote to Forbes in 1849 as "one of the first authorities in England on consumptive cases" (Gaskell 1900, 411), and he was also consulted over Emily's declining health (Gaskell 1900, 390).

"inequality in the functions and powers of the two thinking organs" (1844, 275). Instead of proposing a balance between the two sides, here just one has to take charge and lead the other. The mind is, then, a "discourse" (1844, 27) between the two halves. Insanity comes about not only when both succumb to madness, but also the weaker brain half is "aggravated to defy the control of the other" (1844, 26).

With character developments in *Wuthering Heights* so drastic that many modern readers of the tale find them hard to digest—Heathcliff becomes too dark, Catherine too frail—one should keep in mind that such changes were easily imaginable through the lens Wigan was using. Full of examples of mothers suddenly driven to murder their infants, upstanding citizens turning into sex-crazed monsters, and religious men becoming raving fools, Wigan's often anecdotal narrative provides a multitude of somatic or mental causes[36] for these behaviors. Small brain obstructions, certain bodily ailments, amoral behaviors or thoughts, and even being ever so slightly hit on the head with a ruler (1844, 195) could quickly transform a specific feeling into its polar opposite and honorable men into Gothic lunatics. While a figure like Heathcliff might still act in a fully rational fashion, a minute change in the brain could be enough to account for his deeply irrational hatred for his competitor and his offspring. According to Wigan's theory, the only thing that keeps most people sane is the dual nature of the brain. Should one hemisphere become temporarily insane or fail altogether—and given how frail the mind seems in Wigan's descriptions, this does appear to be a fairly common occurrence—the other, healthy side can take over. As long as the dual brain works together, insanity can (generally) be avoided. But clearly, this is not the case at Wuthering Heights.

Perhaps the most graphic and up-front depiction of flat-out "insanity" is Catherine's breakdown in chapter 12. As she is talking to her servant after having confined herself to her room for several days, we observe her slipping from confusion and outrage into delusion. Beginning with her inability to recognize her reflection in a mirror ("That is the glass—the mirror, Mrs. Linton; and you see yourself in it"), she soon "increase[s] her feverish bewilderment to madness," as our narrator in this situation (the servant Nelly) puts it (Brontë 1848, 109, 107). With Catherine's brain heating up and slipping in and out of consciousness (and in and out of non-/unconsciousness), she loses

---

[36] One of them, a throwback of sorts to heroic medicine, appears to be gout, a hobbyhorse of Wigan's and cause for many of the mental ailments he described.

80   WRITING THE BRAIN

a firm grasp of what her "self" is. Within minutes, Catherine shows signs of an actual fever ("all burning"), experiences mood swings ("changes of her moods"), and slips into a "childish" state (cf. Moglen 1971, 394). When sanity seems to return momentarily, Catherine argues that her "brain got confused" and even discloses a prior state of catalepsy: "Blackness overwhelmed me, and I fell on the floor . . . . I had no command of tongue, or brain" (Brontë 1848, 109).

Although Nelly seems to be accusing Catherine of "faking it" to agitate her husband, there seems to be some slippage here between sanity and insanity that is, at least partially, at Catherine's volition. But, as we know, not for long. Looking at Wigan's analysis of hypochondriasis, we might understand why. As an example, he provides us with the following anecdote:

> A friend of mine in the country, a lady of independent fortune, high mental cultivation, and great vigor of mind apparently in perfect health, was the subject of strange and indescribable internal sensations intensely distressing, and often prophesied that she should die—a prophecy ridiculed by all her acquaintance, as well as by the medical attendant. She bore her incredulity with great complacency, and in the apparent exemption from all bodily ailment, died. (1844, 325)

As a reason for cases such as this, Wigan hypothesizes ailments attacking a part of the extended nervous system that cause it to send faulty impulses,[37] or a diseased cerebrum that is misreading nervous information. Since Wigan knows that the motor system is organized hemispherically (e.g., the left hand is moved by the right side of the brain), he reasons that a fantasized malady might actually correspond to an issue within the respective brain hemisphere. If the "tyranny of the sound ratiocinator over the unsound ratiocinator" (Wigan 1844, 65)—that is, of the healthy over the diseased brain—breaks down, specific delusions in the form of hypochondria can escalate into full-blown insanity (Wigan 1844, 324–25).

Another sign, from Wigan's perspective, that one of Catherine's brains might be diseased and struggling against the control of its healthy counterpart is her lapse into memory. The voice of her ailing cerebrum, speaking "childish[ly]," seems to exist on a different plane: This part of her plays with

---

[37] In this case, the brain would be receiving "disordered impressions" (Wigan 1844, 325) from malfunctioning ganglia in the body that it consciously understands as incorrect.

feathers and thinks itself on a hunting excursion with young Heathcliff. Interspersing his own voice with lengthy quotations from the work of Henry Holland, Wigan argues that each brain can carry on a single, separate train of thought—but "one of [these] generally rests in the memory" (1844, 111). The weaker brain is generally delegated to remembrance, while the stronger one is the active planner and in charge of the other. Catherine's infantile side[38] is, then, a Catherine of the past; one that was suppressed and dominated by a rational side that now, overworked and stressed, has lapsed into inactivity. When Susan R. Gorsky argues that "when Catherine attempts to accede to her socially-approved role, she denies her nature and faces personal disintegration" (1999, 173), she might as well be describing this process. As a woman generally ill-equipped to constantly "fight" her other self and affected by several causal factors for insanity—"fever," "inflammation of meninges," "grief," "starvation," and perhaps even "softening of brain" (Wigan 1844, 339)—she is experiencing the resurfacing of the neglected cerebrum of her youth.

While Catherine would certainly have made a fine addition to Dr. Wigan's collection of anecdotal tales of the dual brain, Heathcliff might arguably be the more spectacular one. His mix of cunning performances, cold-hearted calculation, and sudden outbursts of physical violence would likely be labeled as "sociopathic" today. In psychiatric texts of the nineteenth century, a very similar concept existed, and Wigan labels it at numerous points in his text: "moral insanity." Coined by James Cowles Prichard (cf. Wigan 1844, 281), the term was used to describe a "morbid perversion" of certain emotions or cognitive processes while retaining the ability to think rationally. Wigan provides us with the following explication:

> [Moral insanity leads] to a variety of crime, ranging from the forgery which merely anticipates the arrival of property necessarily coming to the offender . . . to the deliberate murder of a benefactor to hasten the enjoyment of the fortune he has bequeathed. The intellect can find reasons to excuse, to palliate, or even to justify the act. (1844, 281)

We can certainly position Heathcliff on the more extreme end of this spectrum. Realizing that she has been tricked by Heathcliff into marrying her, his wife characterizes this dark figure accordingly: "He's not a human being. . . .

---

[38] A state, according to Wigan, caused either by "folly" or through "forced premature development" (1844, 296).

82 WRITING THE BRAIN

I gave him my heart, and he took and pinched it to death, and flung it back to me. People feel with their hearts" (Brontë 1848, 150). Letting us know that there is something fundamentally wrong with Heathcliff, she identifies a total absence of feeling. The only emotion he seems to be able to experience is rage. Agitating him into a blind frenzy—an act she describes as "pulling out the nerves with red hot pincers"—Isabella manages to get him so "worked up [as] to forget the fiendish prudence he boasted of, and [he] proceeded to murderous violence" (Brontë 1848, 151).

Wigan, unlike many of his colleagues, understands moral insanity as an "early stage" (1844, 281) of a process that often begins in "morose and gloomy thoughts and stern self-control" (1844, 285), then slowly slips into criminal but still rational behavior, and ends "before a lunacy commission" (1844, 281). In the last moments of Heathcliff's life, we can certainly hear echoes of this descent: Seeing ghostly visions of Catherine at night, he dies (perhaps from hypothermia, perhaps from a neurological condition) and is discovered with an expression of "frightful, life-like exultation" on his face (Brontë 1848, 286).

While the ghostly hints of Catherine perhaps best serve to classify the tale as "fantastic," they are not safe from the scalpel of Dr. Wigan. Indeed, "spectral illusions," as he calls them, fall directly within his area of expertise, as he shows by discussing the case of a young man who kept seeing apparitions of his deceased mother: Such spectral sightings constitute "past impressions assuming the character of reality, and producing almost an equal effect on the unhappy possessor of a vivid imagination" (1844, 166). "It is only in the disturbed or deranged state of one brain that these fanciful figures are presented," he argues, claiming that rational thought is possible simultaneously via the second brain, hence differentiating the experience from hallucinations during lunacy (1844, 166). With one mind seeing a figure that is not there and the other one not seeing it, the visual image one experiences is a ghostly projection reflecting the discordant relationship of the two brains' visual apperceptions.

As is the case with Catherine, we discover in Heathcliff's cranium an insane mind struggling with "morbid volitions" and fighting a battle between regression into memory and struggling through the present. With the psyches of the former sibling figures twisted and turned away from their prior selves and into paths they did not choose, their brains have turned against themselves. One side retains a childlike state, and the other strains and ultimately breaks under the pressure to control it. We see Catherine and

SPLIT BRAINS, DOUBLED MINDS   83

Heathcliff transition from early stages of mental ailment (moodiness, infantilism) through hallucinations and spectral visions and ultimately into flat-out madness and death. Although Arthur Wigan had already joined the late Catherine and Heathcliff by the time *Wuthering Heights* was published (he died in December 1847), we can be certain he would have had quite a few things to say about this peculiar psychiatric tale by "Ellis Bell."

On the symbolic plane, Heathcliff and Catherine truly seem to embody Wigan's notions of the dual mind: They play out the "perfect individuality of two brains" (1844, 115) that nonetheless constitute a single psyche. But instead of working together, as was the case when they were children, their dual mind has become disrupted, with "each [brain] carrying on a distinct and separate train of thought" (1844, 108). And both are, in Wigan's phraseology, "insane": Catherine is moody and irrational, Heathcliff largely rational but murderous. The personality they once shared is completely broken apart. As with Jekyll and Hyde, the two brains struggle for dominance, with homicidal insanity and death as the unavoidable result. Here the reason appears to be neglect in childhood, a factor explicitly discussed by Wigan (1844, 353). We can also hear echoes of their story in the psychiatrist's example of two young brothers that were once "incessantly together," roaming the fields hand in hand and experiencing physical distress at being separated (Wigan 1844, 191), until one began to exert "very unreasonable and tyrannical authority over the younger" and attempted to murder his brother with a "carving-knife" (Wigan 1844, 193).

Recasting the basic structure of the hemispheric, doubled brain in Wigan's theories, *Wuthering Heights* revolves around bipartite relationships that work in seemingly perfect harmony until dominance of one over the other is questioned or outside forces break up these hitherto inseparable bonds (cf. Levy 1996, 159). Much in keeping with Wigan's examples, such dissolutions then trigger various "swoon[s]," "illnesse[s]," or "brain fever[s]" that ultimately result in death or prolonged ailments (Brontë 1848, 242, 127, 117). In *Wuthering Heights*, "internal division" (Gorsky 1999, 184) describes both of Brontë's characters as the very structure of the narrative itself. Health, in Brontë's tale as in Wigan's brain theories, is two equal partners working together—and illness is the dissolution of this relationship. Though this does, of course, resonate heavily with Romanticist notions of passionate love between the sexes and/or affection for one's sibling, *Wuthering Heights* seems to go far beyond such notions in the intensity of the relationship depicted—with "incestuous" (Solomon 1959, 83) being only one adjective used to describe it.

84   WRITING THE BRAIN

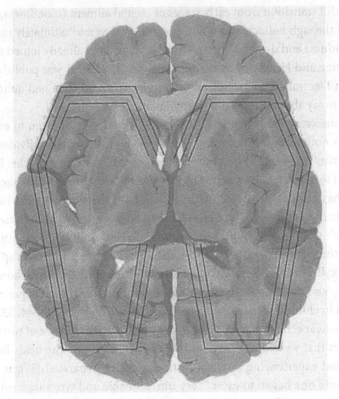

Figure 9 The brain as bicameral coffin.

There is something truly transgressive at play here, perhaps best described by Catherine when she bursts out, "I am Heathcliff! He's always, always in my mind" (Brontë 1848, 72)—a strikingly odd phrasing, even in the nineteenth century.[39] Though by this point of the story these two are almost polar opposites, they still seem to share one skull: Heathcliff is not *on* but *in* Catherine's mind. The former orphan, later in the tale, returns this odd cranial gesture, imagining his corpse buried next to hers, with a breakthrough between the coffins. This vision of a bicameral burial—two persons, each in their box but with a connection in between—is Heathcliff's understanding of a mind at rest: the psyche they shared, now recreated as a deathly cerebrum (Figure 9). With the inner discord of Heathcliff and Catherine's minds

---

[39] Indeed, no identical use of "in my mind" to mean "on my mind" could be located in nineteenth-century lexicons. The *OED* does not list it as an alternative.

echoing the conflict between the two, the Gothic couple appears to become the very expression of insanity: In their clash of "opposing volitions" (Wigan 1844, 262), we see mirrored the discursive turmoil of a moribund mind. *Wuthering Heights* is, indeed, "a stunningly drawn landscape of the mind" (Moglen 1971, 391)—and perhaps even of one, doubled mind.[40]

Master-Minds

Clearly, in the dynamic between Heathcliff and Catherine, the effect of gender can hardly be underestimated. As in most scientific writing of the early nineteenth century, Wigan argues for physiological differences in male and female brains.[41] Seeing many women grow up in a sheltered "moral atmosphere" (Wigan 1844, 420), Wigan claims they are less prone to insanity resulting from a conflict between their two brains—but they are also, therefore, less likely to develop a high "intellect" (1844, 420). But "for men," he claims, "we cannot make a moral atmosphere; they must fight the world as it stands" (1844, 420). Men and women thus experience insanity differently— with swoons and rages as the respective morbid expressions of their upbringing. Still, while these features seem mostly caused by the environment, Wigan does account for masculinized female and feminized male brains, as his discussion of hysteria suggests (1844, 322–21).

In these descriptions of the female as passive, potentially moral, but mentally fragile and the male as conflict-prone and domineering, we can perceive hints of Catherine and Heathcliff, almost identical beings as children but increasingly gendered and pushed apart by their surroundings. In this sense, Wigan's theories confirm Gilbert and Gubar's argument that the novel follows a "once androgynous Heathcliff-and-Catherine [being] conquered by the concerted forces of patriarchy" and divided into two (2000, 274). With one character subject to cruel abuse and neglect and the other educated to become a lady, we know from Wigan (an early believer in neuroplasticity)[42] that their brains are

---

[40] Of course, this notion is not a radically new one. Gilbert and Gubar, most famously perhaps, speak of "alternative sel[ves] or double[s]" in the novel (2000, 265), whereas Helene Moglen sees *Wuthering Heights* exploring "the definition of personality by externalizing and symbolically dramatizing the conflicting elements of a divided self" (1971, 392–93).

[41] See also Rachel Ann Malane's discussion of gendered brains in Charlotte Brontë's works in her *Sex in Mind* (2005, 67–110). Coincidentally, Malane briefly cites Henry Holland in a different context. Wigan is not mentioned her book.

[42] "*I conceive* that it is in the power of almost every man to produce new groupings, combinations, and aggregations [of brain fibers]" (Wigan 1844, 444).

86 WRITING THE BRAIN

physically adapting to accommodate their new social position. The developing disconnect between the former and current self must lead both characters to their demise: one after being bedridden from a mysterious mental ailment (see Wigan 1844, 143), the other after a "whirlwind of . . . rage" (Wigan 1844, 194).

If we look again at how Catherine's and Heathcliff's personalities split after their disastrous excursion to Thrushcross Grange, we discover their oppositional cerebra engaging each other in a fascinatingly dysfunctional dynamic. Increasingly, while Catherine becomes ladylike, passive, and focused on representation, Heathcliff transforms into a dark and brooding rebel figure. Upon realizing their split—arguably at the moment of Catherine's utterance that "it would degrade [her] to marry him" (Brontë 1848, 71)—Heathcliff and Catherine enter into a discursive pattern that defines their relationship and the novel as a whole. From this point in the story, Catherine becomes reactive and her dark doppelgänger completely takes over the plot. From his flight and return to his manipulation of Isabella and the final act of domination over Cathy, Heathcliff becomes the driving force of the narrative, and Catherine (real or ghostly) is completely at the mercy of his deeds. Instead of looking backward to a society and tradition he never belonged to in the first place, the dark figure violently creates radically new paths for himself, embracing capitalism, buying up lands, and persuading or bending to his will the people in his way. In the struggle for cerebral dominance, Heathcliff's side takes charge.

Wigan himself seems to make a societal argument regarding such developments, based on the simple observation that, as with plots, brains at rest are quite boring. Looking at what he sees as a "chain of progression" (1844, 346) that accounts for the development of society as a whole, Wigan claims that only unruly brains, in (slight) discord with themselves, can advance humanity. If both sides of the brain were truly alike, he proposes, there would be no internal discourse and, hence, no "reasoning" can take place. Discourse between one's "Two Voices," ultimately, is a necessary healthy feature of the brain as much as it is a threat to its sanity if taken too far. The doctor, therefore, concludes that with "both brains (or both halves of nervous matter) being exactly equal, [a person] acts from uniform impulse only" (1844, 349) and is incapable of intellect: "The distinction between mind and instinct consists in the parity or disparity of the two cerebra" (Wigan 1844, 347). It is a dynamic, master-slave dialectic, biologized as brain science. And Heathcliff is aware of it. He assures Catherine he would never hurt her: "The tyrant grinds down his slaves and they don't turn against him" (Brontë 1848, 99). While he is in a state of rebellion, his mind is still a slave to her.

To generate progress, Wigan makes clear, requires that opposing volitions "fight" it out (1844, 240). What we can sense through *A New View of Insanity*, then, is a rather complex network of interlocking dual processes: two highly uneven cerebra struggling for dominance at various levels—the levels of the individuals and society, protagonists as well as narrators. To achieve peace of mind, one volition has to dominate the other. Becoming "master of [one's] own actions" means enslaving one's other mind, which stews in resentment just below consciousness. While never going as far as his American colleague Dr. Redfield in flat-out ascribing a gender to each hemisphere, Wigan's competitive cerebra suggest as much: Each man, one might read between the lines, is suppressing a subservient mind that takes on feminine features—and society as a whole is suppressing the subservient mind, while still requiring its perspective as a dialectical antithesis. This master and slave relationship, of course, not only is gendered but also echoes notions of race.

The dormant, "other" self, deeply impressed into the physical brain and forced into submission by the dominant self, had a long history of subtle racialization that was slowly becoming explicit: *Edgar Huntly*'s Clithero, in sleepwalking, completely loses his culturedness and reverts to being a half-naked, savage Scots-Irish (one of the prime racial "others" of the day), and Huntly himself is suddenly endowed with the somnambular ability to navigate caverns used otherwise only by the Lenni Lenape. Even Brontë's Heathcliff is so often described as "dark" and brutish that many critics now read him as a Romany stereotype of the day.[43] Indeed, Wigan himself emphasizes that the brain of the "savage" and the Western man differ not in kind but only in cultivation. If the nobler parts of the psyche are not trained and refined, they atrophy and the base nature of the "human animal" again takes over (1844, 155). "Christophe, the negro ruler of Haiti," Wigan quotes John Barlow (1798–1869), "was probably not removed above a generation or two from the African savage; yet his daughters were polished and accomplished women, fit to take their place in European society" (in Wigan 1844, 400–401). The specter of slave revolt—personified by the figure of Henri Christophe (1767–1820)—here becomes a metaphor for the suppressed other of the human mind, that unrefined, primordial echo of the self, lurking below the threshold of consciousness. In Wigan's Lamarckian worldview,[44]

[43] See Meyer 1996, 153; Bardi 2008; Althubaiti 2015.

[44] Wigan/Barlow clearly disclose a Lamarckian mindset here that claims humans and other animals only differ as a result of the behavior of the ancestors that shaped them. This accelerated view of evolution (as a question of decades, not millennia) disallowed racial theories that claimed inherent, quasi-permanent difference between human "races" (as misreadings of Darwinian evolution

88 WRITING THE BRAIN

there is a shadowy revolutionary hiding in our very skulls, ready to behead the master of the mind whose cultured arrogance has so long kept him mute.

One of the most accomplished fictionalizations from the era of an actual slave revolt, Herman Melville's 1855 "Benito Cereno," features exactly such a specter: slaves taking control of a ship, the name of which invokes Haiti (*San Dominick* / Saint-Domingue), and inverting the power dynamics on board while keeping up a façade of white dominance. While the theme of slavery is explicit, the tale's psychological themes appear equally dominant, causing some critics to go so far as to deny the story's politics altogether—most notoriously Rosalie Feltenstein, who argues that "slavery is not the issue here" (1947, 254). There is a "self-regulating condition of the mind" at play (Fisher 1999, 95), a portrayal, it seems, of "a mind constitutionally incapable" (Bender 1988, 52)—all taking place onboard one of the prime metaphors for "many in one": a ship.[45]

Of course, Melville scholarship has long discussed the author's fascination with character doubling for psychological effect (Marcus 1962, 366). Michael Paul Rogin, for instance, has noted a tendency toward a "twinning of characters" that intensifies in Melville's later fictions, always resulting not in "resolution" but in "splitting of selves" (1985, 159). To Rogin, however, this splitting is expressly political, suggesting, somewhat in passing, that in Benito Cereno, "Melville made the Spanish captain's 'own shadow' into slavery," a shadow that clings to him in the form of a racialized, doubled Other (1985, 217).

---

later did) but instead biologized imperialist notions of cultural supremacy of the West as a means of perfecting less developed "races." As the encyclopedia *Race and Racism in the United States* puts it: "Lamarck did not emphasize racial differences or hierarchies," and his theories on the transmission of acquired traits became "not as significant in shaping racial attitudes as natural selection" (Laster 2014, 683). Darwin, whose theories later became a driving force for racist science, of course, objected to such appropriation. An outspoken abolitionist (cf. Desmond and Moore 2009), Darwin specifically argued against cranial measurement as a means to "prove" mental interiority of Black populations, observing instead that even *if* these results were accurate (i.e., if Black people's brains *were* indeed smaller), they proved the negative effects of slavery, neglect, and maltreatment on the human mind, not an inherent, racial difference (cf. Darwin 1871, 145).

[45] Think: ship of state, ship of Theseus, or the frequently used "captain of the ship" metaphor for notions of autonomy and (self-)control. All seem to come together as well in Melville's Ahab: "That certain sultanism of his brain . . . became incarnate in an irresistible dictatorship. For be a man's intellectual superiority what it will, it can never assume the practical, available supremacy over other men" (2011, 190). Ahab's domination over his self is what he attempts to transpose onto his ship. Qualities of mind ("sultanism of his brain") become the sociology of the vessel ("dictatorship")—but, as the story discloses, a single element cannot long hold undisputed supremacy over its subservient elements. Not in brains, not on ships.

SPLIT BRAINS, DOUBLED MINDS 89

This shadow looms large, as soon as the tale's free-indirect narrator, Amasa Delano, sets foot on deck:

> Whether the ship had a figure-head, or only a plain beak, was not quite certain, owing to canvas wrapped about that part. . . . Rudely painted or chalked, as in a sailor freak, along the forward side of a sort of pedestal below the canvas, was the sentence, "Seguid vuestro jefe" (follow your leader); while upon the tarnished head-boards, near by, appeared, in stately capitals, once gilt, the ship's name, "SAN DOMINICK," each letter streakingly corroded with tricklings of copper-spike rust; while, like mourning weeds, dark festoons of sea-grass slimily swept to and fro over the name, with every hearse-like roll of the hull. (Melville 1986, 156)

There are a number of themes masterfully at play in this passage—blackness versus whiteness, a burial, and a whipping scene—but also something quite a bit *headier*: a hidden "figure-head," "tarnished head-boards," and a mysterious instruction to follow "vuestro jefe," which also translates to "your head,"[46] all suggesting some cranial trouble. And even before we learn that the canvas indeed hides the skeleton of the dead "master" Don Aranda (killed by his slave Babo), "*stately capitals*"[47] hint at a "noble head" on display beneath.

Ultimately, the story will reveal that the slave upheaval suggested by in this passage had already happened, with the performance of the status quo a ruse by the former slaves to ensure their safe escape. The inversion of the nautical master-slave dynamic, internally and externally (i.e., in the mind and onboard), is embodied by the duo of (former) captain Benito Cereno and the Senegalese slave Babo. Their duality is perhaps best embodied by the notorious shaving scene that turns the coerced submission of tending one's master into an echo of past violence and a threat of future beheading. Here, too, is the first moment where the narrator senses a clear inversion: Just for a moment, Delano sees "a headsman" in Babo, a term denoting his status as both executioner and ruler,[48] a man who takes craniums and overtakes them. In the scene, Cereno seems to attempt to beg Babo for his life while keeping

[46] "Jefe, *sm*. Chief, head, superior, leader" (Seoane 1854, 518).

[47] Etymologically, "capital" stems from "caput" (head) (Harper n.d., s.v. "capital") and "stately" means "noble" in a "sense of costly and imposing display" (Harper n.d., s.v. "stately").

[48] Headsman (n.), "executioner," c. 1600, from genitive of *head* (n.) + *man* (n.). Used earlier in the sense of "chief, leader," c. 1400 (Harper n.d., s.v. "headsman") .

90 WRITING THE BRAIN

up his performance as captain. To this, the slave-turned-master only replies equally performatively: "Now, master" (Melville 1986, 215). The moment is drawn out to an excruciating length, with Babo handling the other man's head so vigorously that the observer for a moment conceptualizes it as his creation: "The negro seemed a Nubian sculptor finishing off[49] a white statue-head" (Melville 1986, 217). Albeit briefly, the slave now has fully turned subject in Delano's mind—grammatically, politically, psychologically—and his master descends into mere objecthood. Before, slaves had been reduced to "whispering" (Melville 1986, 184) on deck, "barren voices" (as Tennyson may have put it) heard only at night. Now, Babo is in command of the ability to speak and Cereno's timid "husky whisper" (Melville 1986, 169) is only to be uttered at his command. The rebellious, shackled Black voice now has taken over not only the ship but Cereno's skull.

Still, Babo's rebellion is doomed to failure, a failure Melville frames as an overreliance on brain:

As for the black—whose brain, not body, had schemed and led the revolt, with the plot—his slight frame, inadequate to that which it held, had at once yielded to the superior muscular strength of his captor, in the boat. Seeing all was over, he uttered no sound, and could not be forced to. . . . During the passage, Don Benito did not visit him. Nor then, nor at any time after, would he look at him. Before the tribunal he refused. When pressed by the judges he fainted. On the testimony of the sailors alone rested the legal identity of Babo.

Some months after, dragged to the gibbet at the tail of a mule, the black met his voiceless end. The body was burned to ashes; but for many days, the head, that hive of subtlety, fixed on a pole in the Plaza, met, unabashed, the gaze of the whites; and across the Plaza looked towards St. Bartholomew's church, in whose vaults slept then, as now, the recovered bones of Aranda. (Melville 1986, 258)

Babo is the head, Babo is the brain. No wonder Cereno, having lost control of these very features, surrenders consciousness when faced with ascribing an identity to the slave by naming him. Cereno's self is so closely mixed with Babo's that differentiating between the two breaks his mind. While Cereno

---

[49] The murderous pun—i.e., another form of "finishing off"—is, of course, intentional. This meaning of "finish" is recorded as early as 1755 (Harper n.d., s.v. "finish").

is nominally in charge again of brain and speech, the effects of losing control over both linger.

Consequently, Cereno, too, has to die once Babo is martyred. In a scene clearly planting the seeds for Conrad's *Heart of Darkness*, the melancholy captain/captive is interrogated about the nature of his depression of spirits. Turning to Delano, Cereno observes:

> "[Y]ou were with me all day; stood with me, sat with me, talked with me, looked at me, ate with me, drank with me; and yet, your last act was to clutch for a monster, not only an innocent man, but the most pitiable of all men. To such degree may malign machinations and deceptions impose. So far may even the best man err, in judging the conduct of one with the recesses of whose condition he is not acquainted. But you were forced to it; and you were in time undeceived. Would that, in both respects, it was so ever, and with all men."
>
> "You generalize, Don Benito; and mournfully enough. But the past is passed; why moralize upon it? Forget it. See, yon bright sun has forgotten it all, and the blue sea, and the blue sky; these have turned over new leaves."
>
> "Because they have no memory," he dejectedly replied; "because they are not human."
>
> "But these mild trades that now fan your cheek, do they not come with a human-like healing to you? Warm friends, steadfast friends are the trades."
>
> "With their steadfastness they but waft me to my tomb, Señor," was the foreboding response.
>
> "You are saved," cried Captain Delano, more and more astonished and pained; "you are saved: what has cast such a shadow upon you?"
>
> "The negro." (Melville 1986, 257)

The forces of nature caressing Cereno's face only echo a different "nature": Babo, his inverted other, slicing his cheek with a razor. Having been forced into the position of "the negro," Cereno found Babo within him: his fear, his humiliation, his hatred. This inner duality, his exercise in forced empathy made him realize, is a key feature of the human mind. There exists a gulf between self and other—indeed, between self and self—deep in "the recesses" of one's "condition" that Cereno can never adequately verbalize. There is indeed a "chasm between [his natures] so profound and obscure,

92   WRITING THE BRAIN

that though perpetually traversing it in all the functions of life," men like Delano have yet "no eye to penetrate its depth" (Holland 1839, 178). "Three months after being dismissed by the court," Melville concludes his tale, "Benito Cereno, borne on the bier, did, indeed, follow his leader" (1986, 258). Who this leader is, Melville underscores by triangulating the location of Cereno's death: in broad view of Babo's proud head, glaring at the burial place of his former "master" Aranda, hidden deep below a church named for St. Bartholomew, the beheaded apostle who was once sold as a slave. Another inversion, it seems.

While Cereno has to learn what it is to be a slave (a lesson that ultimately kills him), the "hive of subtlety" of a subjectivity forced into objecthood for too long has developed a keen sense for the mind of masters. Babo's plan to not attack Delano's boarding party but force the remaining white crew into a painful theatrical performance to trick his visitors almost succeeded, and would have had it not been for Cereno's desperate attempt at flight. The white supremacy engrained as natural into Delano's mind (allowing him to explain away most of the dissonance he experiences onboard) Babo understands as an implicit cognitive theory and weaponizes it against his aggressors. His ability to hold such widely contradictory systems of thought—the two voices of slave *and* master—in his head simultaneously has led several critics to read the tale through W. E. B. Du Bois's notion of "double consciousness." The mental strategies of Babo have been cast as a "parodic inversion" of Du Bois's later theocratization of the Black experience (Siemerling 2005, 80; cf. Sullivan and Tuana 2007, 10), one "similar in one respect but paradoxical in other respects" (Rampersad 1997, 171).

Where did Du Bois learn about the term "double consciousness"? A rather persuasive link is William James (1842–1910), that leading early voice of modern psychology and Du Bois's professor at Harvard. There has been substantial scholarly argument about congruences and divergences between the self-described "devoted follower of James" (Moses 2004, 213) and his mentor, who was then lecturing from what would become his groundbreaking *Principles of Psychology*.[50] Still, at least some of the caution about assuming too close a connection is based on an error. The *Stanford Encyclopedia of Philosophy* summarizes the scholarly consensus: "James does not seem to have used the term 'double-consciousness'" (Pittman 2006). That is incorrect. While volume 1 of *Principles* does not mention the term, we can

---

[50] See, for instance, Bruce 1992; Posnock 1997; Reed 1997, 100–112.

find it invoked in volume 2 (James 1981, 1200), in a chapter on a familiar theme: hypnotism, sleepwalking, trance. While James's sourcing is rather lax in this part of the book (he does not source the term itself, for instance), we find its history permeating the margins of *Principles*, from its repeated references to Wigan's theories on the duality of mind (James 1981, 369, 635–36, 1170–71), right back to the very book that defined double consciousness so tantalizingly at the beginning of this chapter: Henry Holland's 1839 *Mental Notes* (James 1981, 199, 1337).[51]

The long, largely forgotten history of the dual mind, then, may live on even in contemporary, subtler notions of double consciousness as a faint, shadowy doppelgänger, a determinist terror haunting the material mind. Its echoes harken back through Rush's musings on sleepwalking embodied by Edgar Huntly's two minds torn between the colonial project and "going native"; its harsh winds are bellowing through *Wuthering Heights*; it is at the root of Tennyson's doubled embrace of the beauties of death and the marvels of life that may never reach synthesis; it shatters Cereno when he is forced to inhabit the mind of a slave and that of a master of slaves simultaneously. Each invocation of the term discharges a minute spark of Holland's perplexed awe over a "consciousness, [that] makes us aware of unceasing change, [but] tells of no breach of continuity" to the healthy self (1839, 173)—an awe over a mind that may not be as unitary, coherent, and logical as it tells itself. The "doubleness" of the "cerebral structure" surely must have profound "effects on . . . the general economy of life" (1839, 178), Holland concludes in his chapter on the dual brain. From the relative distance of a decade of "two Obamas" (Alter 2013), "two different Donald Trumps" (Petri 2016), and "two Joe Bidens" (Tumulty 2020), each contained within a more or less functional single cerebrum, we may agree.

---

[51] The specific citation is to Holland's 1852 *Chapters on Mental Physiology*, an expanded edition of his 1839 *Mental Notes*, which includes his chapter on the double brain and his definition of double consciousness. It should also be noted that Ralph Waldo Emerson, who also defines the term in his essay "Fate," figures as another key source for Du Bois. While it is unclear when Emerson learned about Henry Holland, the two did meet in 1873, shortly before Holland's death, in London.

3

Skulls and Society

Reading the Mind as a Multi-Organ Entity

Brain of the New World, what a task is thine,
To formulate the Modern—out of the peerless grandeur of the
   modern,
Out of thyself, comprising science, to recast poems, churches, art
   —Walt Whitman, "Thou Mother with Thy Equal Brood"

A curious sect's in vogue, who deem the soul
Of man is legible upon his poll:
Give them a squint at yonder doctor's pate,
And they'll soon tell you why he dines on plate
   —Charles Tennyson Tuner, "Phrenology"

Brain Damage

On September 136, 1848, Phineas Gage was leaning over a stretch of rock
outside of Cavendish, Vermont, to set up an explosive charge to clear the way
for the Rutland and Burlington Railroad. He had already drilled a deep hole,
inserted the blasting powder, and was stuffing it down with a long iron rod
when he heard a voice from behind. He turned around to face the colleague
who was addressing him when something went terribly wrong. Gage probably did not even register the sound of the accidental detonation before the
almost-four-foot-long rod (see Figure 10) shot out of the barrel-like hole and
went straight through his head, landing several feet behind him.

His coworkers rushed Gage to a local doctor, John Harlow, who later recorded his impressions of the patient:

Writing the Brain. Stefan Schöberlein, Oxford University Press. © Oxford University Press 2023.
DOI: 10.1093/oso/9780197693681.003.0004

Figure 10 Phineas Gage posing with his iron rod (originally from the collection of Jack and Beverly Wilgus, and now in the Warren Anatomical Museum, Harvard Medical School).

> From the appearance of the wound in the top of his head, the fragments of bone being lifted up, the brain protruding from the opening and hanging in shreds upon the hair ... The globe of the left eye was protruded from its orbit by one-half its diameter, and the left side of the face was more prominent than the right side. The pulsations of the brain were distinctly seen and felt.
>
> The scalp was shaven, the coagula removed, with three small triangular pieces of the frontal bone, and in searching to ascertain if there were foreign bodies in the brain, I passed the index finger of the right hand into the opening its entire length, in the direction of the wound in the cheek, which received the left index finger in like manner, the introduction of the finger into the brain being scarcely felt. (1869, 6–7)

96 WRITING THE BRAIN

Much to the surprise of everyone involved, especially Harlow, Gage survived the ordeal. Even more surprisingly, he seemed fine, except for some scars and a blind left eye. Being pierced through the brain with a thick iron spear seemed to leave the man rather unaffected, and Gage did not seem to miss the significant amounts of brain the doctor had removed: At first glance, he moved and conversed like a healthy human being.

Once Harlow published his first paper on the surprising case, Gage quickly became a minor celebrity, posing with the rod at Barnum's, and being discussed in newspapers from New York to Wisconsin and even England.[1] To this day, he remains one of the most frequently discussed medical cases of the nineteenth century.[2] Still, what makes his case special and an enduring part of the *Aufschreibesystem* brain, is less his fascinating recovery than what happened to Gage as a result of the accident: His personality changed. Once a hardworking man of common sense and neighborliness, as Harlow described him, Gage morphed into the opposite:

> He is fitful, irreverent, indulging at times in the grossest profanity (which was not previously his custom), manifesting but little deference for his fellows, impatient of restraint or advice when it conflicts with his desires, at times pertinaciously obstinate, yet capricious and vacillating, devising many plans of future operations, which are no sooner arranged than they are abandoned in turn for others appearing more feasible. (Harlow 1869, 13)

Whether or not we want to follow the local surgeon in the extent of his assessment, the still-famous Gage case is often "judged to be the first and possibly most important [case] to reveal something of the relation between the brain and complex personality characteristics" (Macmillan 2002, 1)—in particular, the relationship between planning abilities, affect control, and the prefrontal cortex (García-Molina 2002).

While the Gage case certainly was not the first of its kind—that appears to be E. de Nobele's 1835 description of the personality changes of a youthful suicide survivor—the spectacular nature of the accident made it the most widely reported. Still, in many popular retellings of this story today, the Gage

---

[1] See "Horrible Accident" 1848; "A Strange Case" 1848; and "Extraordinary Accident and Cure" 1850.

[2] On Gage's life and legacy, see Damasio et al. 1994; Fleischman 2010; Macmillan 2002; and perhaps even Glass 2006.

case has morphed into an even more problematic kind of "first"—namely, the "first" hint of the "brain's role in determining personality" (as the Wikipedia entry on the case currently puts it). This assessment is certainly false. Indeed, by the time Gage began his shift on that day in September 1848, the effect of brain damage on personality and experience of self was as well known as it was little understood. Indeed, it had become such a household truism that brain damage became a recurring theme in the literature of the day. From now canonical works to obscure periodical fiction, a good blow to the head suddenly went a long way in creating an intriguing plot.

Akin to the common trope of the amnesiac protagonist in the late twentieth century in film and television,[3] many prose works in the early to mid-nineteenth century relied on traumatic brain injury as a key moment of character development. There is, for instance, the protagonist of Walt Whitman's early novella fragment "The Fireman's Dream" (1844), who needed "a tremendous blow on his head" to connect to his mysterious past and develop visions of the future; the eponymous protagonist of George Eliot's *Adam Bede* (1859), who has to supply "a well planted blow" to the skull of his romantic rival (254), causing unconsciousness and a change of heart; even the sister of Charles Dickens's Pip, who is "knocked down by a tremendous blow on the back of the head" (1881, 143), as a result of which Pip finds the "temper" of this once hard and reproachful woman "greatly improved" and her whole demeanor now one of "patience" (1881, 146).[4] A lesser-known tale like "Remembrances of a Monthly Nurse" by Harriet Downing, published in 1838 in the British *Fraser's Magazine*, even goes so far as describing quite graphically the effects of "a violent concussion on the brain" of a young "imbecile," who has his skull kicked in by a horse and loses brain tissue ("a large quantity of the actual brain exclude[d] from the fracture . . . as much as a hen's egg"), and as a result turns into a genius and is fully healed of his mental handicap (1837, 504-5).[5]

---

[3] See "TV Tropes" n.d.

[4] A more toned-down version of this trope is the still-common narrative shortcut of a character finding themselves in a situation said person would never have entered voluntarily (i.e., it would be out of character) but unconsciousness following head trauma has placed him or her there. The young-adult novellas of Horatio Alger (1832–1899) feature such moments frequently—for instance, in his *Randy of the River* or *Nelson the Newsboy*. Whitman also employs this version of the trope in *Jack Engle*, where a head injury leads the eponymous narrator to meet his future wife when her (foster) mother bandages his injured friend's head and hides "phrenological developments from the public gaze" with a handkerchief (2017, 111).

[5] The piece even includes a lengthy footnote explaining the medical science behind said "true case": "The centre of the brain's vitality appears to be at the base, where the fibres proceeding from the spinal marrow, which enter into the structure of its hemispheres, cross and recross each other;

98 WRITING THE BRAIN

Besides the character changes such blows to the head bring about, these moments all share specific characteristics. Narratively, they not only involve the appearance of a surgeon but also describe in great detail the immediate somatic effects of a severe concussion to the head: language loss, loss of balance and consciousness, breathing problems, loss of vision, fevers and shaking, a lengthy recovery period, and the potential for permanent damage (cf. Chelius 1847, 410–11; Solly 1848, 316–19). With such thorough and precise medical information condensed into these short moments, why do their narrative effects, then, seem so out of the ordinary (at least to readers today)?

Perhaps one of the more famous, and certainly one of the earliest, of these literary head injury cases can begin to provide an answer: the famous accident at Lyme in Jane Austen's *Persuasion* (1817; see Figure 11). It features a character by the name of Louisa Musgrove attempting to jump feet-first down a harbor wall into a man's outstretched arms. He, however, fails to catch Louisa in time and her head hits the pavement, leading to immediate unconsciousness, problems breathing, a long period of recovery—and a change of mind: "The day at Lyme, the fall from the Cobb," the narrator later muses, "might influence her health, her nerves, her courage, her character to the end of her life" (Austen 1899, 216). What would today be considered a traumatic brain injury has had a quite dramatic influence on Louisa: As a result of her fall, she suddenly turns into "a person of literary taste, and sentimental reflection" (Austen 1899, 216), qualities much at odds with her former personality. Alan Richardson has rightly described the passage as an important example of the "embodied mind" (2001, 99) in Austen and a reminder of the political and religious debates of her time surrounding mind-materialism (Richardson 2001, 99–110). Still, Richardson does not further elaborate on why, exactly, Louisa's "headfirst fall onto the paving stones" (Richardson 2001, 109) might have caused such a change.

One branch of brain science in Austen's time would have provided a rather clear-cut answer to this abrupt cranial readjustment: the *Schädellehre* of Franz Joseph Gall (1758–1828) and Johann Gaspar Spurzheim (1776–1832). A "headfirst fall" would have meant hitting the area of the craniological brain

---

and this part cut or punctured, death instantly ensues. This constitutes what is generally known as the *pithing operation*, by which all animals are instantly killed. The brain, according to phrenologists, is a congeries of organs, each of which, from the top of the spinal marrow, sprouts up in the form of a cone, which extends its base under the internal surface of the skull. Hence, it may easily be understood, that if, phrenologically, the cerebral substance composing the organs be collectively so much 'confined, cribbed in, and cabined,' so that they cannot individually exercise their activity, the removal of a portion of brain, even by accident, might relieve the oppression, and they might find 'ample verge enough and space' to use their wonted functions" (Downing 1837, 506).

Figure 11 Louisa's fall, 1889 illustration by C. E. Brock (Austen 1906, 157).

associated with the faculty of "Individuality,"[6] located behind the center of the eyebrows (Gall and Spurzheim 1815, 181). This region, craniologists believed, was used to "acquire knowledge of the external world" (Gall and Spurzheim 1815, 181). The stronger this faculty, the more matter-of-factly a person was said to think; a less active (or damaged) Individuality would thus lead to introspection and musing (Gall and Spurzheim 1815, 181). Some later craniologists even went as far as to claim that if people with a highly developed Individuality read at all, they gravitated toward real-life subjects

---

[6] On the following pages, phrenological terms are capitalized to differentiate them from common use.

100 WRITING THE BRAIN

such as biography or history (Morgan 1871, 324). A worldly figure like Louisa could thus, with a good blow to her Individuality, be turned inward, her obsession with the mundane morphing into "sentimental reflection" (Austen 1899, 216).

While Austen's exact knowledge of the science remains unclear, the 1810s saw craniology/phrenology slowly becoming a mass phenomenon, first sweeping the medical journals of Europe and the United States (as early as 1803) and then turning into a popular trend that lasted through the rest of the century. Although its later life at the nexus of fortune-telling and pseudoscience might have soured appreciation of the theory today, the teachings of Spurzheim and Gall in their day were not only a cultural force to be reckoned with but also a milestone in thinking about the brain.

Indeed, the fundamental basis of the teachings of Gall and his devoted student Spurzheim are now rightly considered "the first [theory of the] cerebral cortex [as] composed of functionally distinct areas," an idea that has become a staple of modern neuroscience and one that had "not [been] given serious consideration" until the early 1800s (Finger 1994, 32).[7] Believing (incorrectly) that "from the beginning of ossification till death[,] the internal table of the skull is moulded according to the brain" (Gall and Spurzheim 1815, 35), these two leading Austrian anatomists surveyed around 450 heads to conclude that specific behavioral traits, emotions, and temperaments seemed to correlate with increases or decreases in the thickness and shape of certain regions of the skull ("bumps" and "depressions"). Gall originally postulated that twenty-seven faculties could be distinguished thusly (the aforementioned "Individuality" being just one), a number that Spurzheim later increased slightly (Finger 1994, 33). Scandalously, the vast majority of these faculties (nineteen, according to Gall) were believed to be shared between humans and (other) animals. This fairly heretical statement resulted in Gall being banned from public speaking in Vienna, forcing him to continue his research in Paris.

---

[7] Boshears and Whitaker perhaps provide the most succinct summary of the value of phrenology: "The supporters of phrenology correctly argued that the key elements of character and personality were located in the brain; they dubiously argued that these were reflected in detectable protrusions or indentations of the skull. The detractors of phrenology correctly argued that the physical evidence, the bumps and dents in the skull, did not reliably correlate with character and personality traits; they dubiously rejected the notion that traits might be localizable in brain regions" (2013, 88). The basic phrenological tenant of brain localization is still so prevalent that neuroscientific polemics urging a break with the idea still frequently invoke it in their titles—for instance, Michael L. Anderson in his *After Phrenology: Neural Reuse and the Interactive Brain* (2021), or William R. Uttal in his *The New Phrenology: The Limits of Localizing Cognitive Processes in the Brain* (2003).

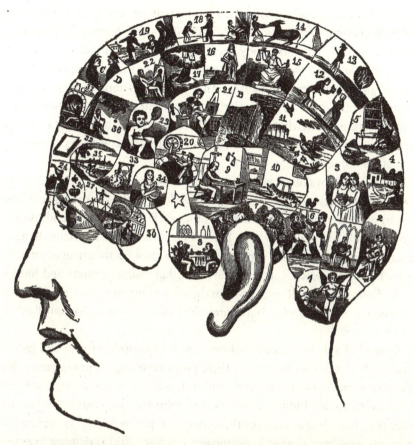

NUMBERING AND DEFINITION OF THE ORGANS.

1. AMATIVENESS, Sexual and connubial love.
2. PHILOPROGENITIVENESS, Parental love.
3. ADHESIVENESS, Friendship—sociability.
A. UNION FOR LIFE, Love of one only.
4. INHABITIVENESS, Love of home.
5. CONTINUITY, One thing at a time.
6. COMBATIVENESS, Resistance—defence.
7. DESTRUCTIVENESS, Executiveness-force.
8. ALIMENTIVENESS, Appetite, hunger.
9. ACQUISITIVENESS, Accumulation.
10. SECRETIVENESS, Policy—management.
11. CAUTIOUSNESS, Prudence, provision.
12. APPROBATIVENESS, Ambition—display.
13. SELF-ESTEEM, Self-respect—dignity.
14. FIRMNESS, Decision—perseverance.
15. CONSCIENTIOUSNESS, Justice—equity.
16. HOPE, Expectation—enterprise.
17. SPIRITUALITY, Intuition--spiritual revery.
18. VENERATION, Devotion—respect.
19. BENEVOLENCE, Kindness—goodness.
20. CONSTRUCTIVNESS, Mechanical ingenuity.
21. IDEALITY, Refinement—taste—purity.
B. SUBLIMITY, Love of grandeur.
22. IMITATION, Copying—patterning.
23. MIRTHFULNESS, Jocoseness—wit—fun.
24. INDIVIDUALITY, Observation.
25. FORM, Recollection of shape.
26. SIZE, Measuring by the eye.
27. WEIGHT, Balancing—climbing.
28. COLOR, Judgment of colors.
29. ORDER, Method—system—arrangement.
30. CALCULATION, Mental arithmetic.
31. LOCALITY, Recollection of places.
32. EVENTUALITY, Memory of facts.
33. TIME, Cognizance of duration.
34. TUNE, Music—melody by ear.
35. LANGUAGE, Expression of ideas.
36. CAUSALITY, Applying causes to effects.
37. COMPARISON, inductive reasoning.
C. HUMAN NATURE, perception of motives.
D. AGREEABLENESS, Pleasantness—suavity

Figure 12   Phrenological chart used by the Fowler brothers (Fowler 1847, 4).

102 WRITING THE BRAIN

The theories of Gall and his followers were undeniably troubling to many of their contemporaries. Not only did the two argue against the grain of a dominant belief in mind-body dualism by turning "mind" into *observable physiology* but they also cast some doubt on the still lingering belief in enlightenment-era rationalism. After all, if emotions, tendencies, and temperaments could be particularized into distinct structures within the brain, working somewhat independently of each other, how could a person truly claim to observe and emote in any sort of appropriate, correct, response to outside stimuli?

Indeed, the physiological basis of rational thought is one of the few cases where the two craniologists disagreed: Gall flatly denies the existence of any organ of judgment but instead argues that each faculty could tend toward good or bad (Gall and Spurzheim 1815, 127), while Spurzheim claims to have found enough evidence to stipulate that there must be an "organ of justice," a particular biological center in the brain that weighs options and judges right from wrong. Still, even to Spurzheim, this organ cannot act alone in decision-making processes but is enmeshed with and dependent on all other brain regions.

Instead of a divine voice of reason, able to *objectively* observe and judge, "rationality" was transformed by these proto-neurologists into a messy debate between dozens of distinct, semi-independent neural centers. It is no wonder, then, that Phineas Gage's doctor, John Harlow, would turn toward this exact branch of science (by then known as "phrenology")[8] to explain his patient's wondrous change in personality (Barker 1995) as damage to a singular "organ" of his brain.[9] Under the anatomical gaze of Gall and Spurzheim, the mind thus ceased to be a unitary, indivisible entity and instead had become a multifocal set of organs—a system of ridges and grooves, observable from the outside and manipulatable using brute force.

Still, why would a comparatively crude narrative instrument like head trauma—a deus ex machina more reminiscent of Thor's hammer than of a

---

[8] Spurzheim favored and promoted the term, while Gall had his reservations, given its etymological focus on mind (*phreno*) over skull or brain.

[9] Having studied with a leading US phrenologist and being a lecturer on phrenology himself, Harlow went out of his way to adjust Gage's case to fit phrenological theses. As Fred Barker II puts it: "[If] Harlow consulted the diagram of Gall's phrenological system in [his professor's] physiology textbook, he would have found several mental functions in the path of Gage's tamping iron: poetical and musical talent, language, color discrimination, acquisitiveness, 'comparative sagacity', and the sense of relations of numbers. At a loss to demonstrate any musical, poetical, or language deficiency in this patient, Harlow interpreted Gage's mental imbalance as a disorder of 'comparison and number', that is, an inability to estimate monetary worth or size" (1995, 676).

crane-lifted Athena—rise to such prominence with authors writing during the height of the phrenology craze? Cognitive neuroscientists Rhonda Boshears and Harry Whitaker provide the first hint, dropped rather non-chalantly into the midst of their overview of phrenology in Victorian literature: "Phrenology did not allow for free will" (2013, 97). This certainly posed a problem to the proto-Realists and Realists so infatuated with the science, be it Eliot (Boshears and Whitaker 2013, 100–103; Claggett 2011), Whitman (Mackey 2005; Borst 2014), or the ubiquitous Dickens (Boshears and Whitaker 2013, 96–98). How could a genre that understood itself as socially conscious and partial to the poor, and that seemed obsessed with rags-to-riches plots disallow or bracket free will?

An answer to that question lies in the much-discussed shortcoming of (early) literary Realism in moving from description to proscription—in its inability to write the Real *as it is* without denying the possibility that it could be different. Focusing specifically on the political advocacy of Realism, Amy Kaplan frames this dilemma perhaps most succinctly: "Realistic novels have trouble ending because they pose problems they cannot solve, problems that stem from their attempt to imagine and contain social change" (1988, 160). In writing the problem, Realism reinforces it, sanctifying it with its elegant prose. Ultimately, many have noted, Realism proposes stasis: Social mobility is reserved for characters out of their *proper* place. Whitman's Jack Engle, for instance, has been rich all along and his ultimate victory is only a realization of this truth. The same holds true for the Pips and Twists of Dickens and Horatio Alger's street urchins, and even works like *Middlemarch* or *The Rise of Silas Lapham* end on a note of social inertia. Phrenology supplied to this stasis a psychology: a brain-centric concept of the mind that in its oft-repeated mantra "know thyself" also always suggested "know thy place." Mobility here was reserved for those out of place—and a true change of mind necessitated a good knock to the brain.

Phrenology as Sociology

At first glance, it seems counterintuitive that such a highly contrarian scientific theory of brain function proposed by two Austrian anatomists could become a cultural force powerful enough in the United Kingdom and United States that one could argue it had a structuring effect on a major subgenre of literature. While Gall and Spurzheim might have managed to raise some

104   WRITING THE BRAIN

eyebrows with the occasional well-read or medically inclined reader, they did need some cultural midwifery to catapult them into omnipresence—and it arrived in the form of 1828's *The Constitution of Man*, a detailed, highly readable introduction of phrenology to the masses and an attempt to glean a positive societal vision from a seemingly all too deterministic outlook on self and psyche.

Written by George Combe (1788–1858), a well-versed lawyer who turned phrenology advocate after seeing Gall dissect a brain in 1815, *The Constitution of Man* quickly became one of the century's bestsellers, with close to half a million sales worldwide during the 1800s (Wright 2005), easily outperforming books like *A Tale of Two Cities*, *Great Expectations*, and *A Christmas Carol* during Dickens's lifetime (Patten 2017) or *The Scarlet Letter* during Hawthorne's (Winship 2001). Its total sales in the United States even rival those of *Uncle Tom's Cabin* in the years leading up to the Civil War (Winship 2001). If Lincoln could, tongue in cheek, credit Harriet Beecher Stowe's book with starting a war, one should certainly not underestimate the power of Combe's contemporaneous bestseller.[10]

Perhaps surprisingly, given its focus on brain anatomy, *The Constitution of Man* understands itself primarily as a work of social reform. "At present," Combe observes therein, "the animal propensities are fortified in the strong intrenchments of social institutions: Acquisitiveness, for example, is protected and fostered by . . . accumulating wealth, . . . pride and vanity, by our artificial distinctions of rank and fashion, [etc.]." But by "teaching mankind the philosophy of their own nature . . . they may be induced to modify [them]" (1841, 393). This points directly to the central conundrum of phrenology: How can one modify one's "own nature"? This is precisely what Combe's "moral and political philosophy" (1841, ix) sets out to answer.

While Combe was personally highly skeptical of (if not flat-out opposed to) metaphysical concepts such as "free will," spirit, or soul (Grant 1965, 146), the reasoning he set out in *The Constitution of Man* was decidedly more strategic. As A. Cameron Grant notes: "Physiological determinism and moral responsibility, for so long the battlefield of warring factions, was dealt with by Combe with consummate brevity; Will and Responsibility are merely terms descriptive of human behavior" (1965, 142). Instead of engaging in a line of reasoning that would lead directly into the highly unpopular realm

---

[10] Beecher Stowe, of course, was also well read in phrenology and included elements of it in *Uncle Tom's Cabin*. See Fletcher 2015.

SKULLS AND SOCIETY   105

of "Materialism, Fatalism, and Infidelity" (Combe 1841, 351), Combe talked about fulfilling one's biological *potential*, a strategy that anticipated, some might even say "*inaugurated*[,] the self-help movement" (van Wyhe 2004, 189). Instead of celebrating or bemoaning the limits of human agency, Combe presupposes them—illustrated at times by examples that directly ridicule "free will" (1841, 263)—and instead "advocate[s] personal responsibility and the possibility of naturally sanctioned self-improvement through education or proper self-control" (van Wyhe 2004, 189).

This idea grows out of fertile ground prepared by Gall and Spurzheim. Based on their notion that each mental organ can act appropriately or inappropriately—can be good or bad—Combe argues that the "operations [of each faculty] are right only when they act in harmony with each other" (1841, 58). Instead of suppressing urges, each person must find "gratification [that] is legitimate and proper" for a given propensity (Combe 1841, 54). In an ideal phrenological society, a specific constitution thus feeds perfectly into one's lot in life. Social status, form of employment, and number of children, for example, could be looked at as either a natural fit or an abuse of one's biological potential. Each position comes with a set of biopsychological rewards and penalties, now no longer justified by a divine overseer but by natural law:

> The laborer, for example, digs the ground, and the squire engages in the chase; both pursuits exercise the body. The penalty of neglecting this law is imperfect digestion and disturbed sleep, debility, bodily and mental; lassitude, and, if carried to a certain length, confirmed bad health and death. The penalty for overexerting these systems is exhaustion, mental incapacity, the desire of strong stimulants, such as ardent spirits, general insensibility, grossness of feeling and perception, with disease and shortened life. (Combe 1841, 49)

By applying Combe, the upper strata of society could justify their position as a proper result of their inherent constitution, while the educated middle and lower classes could justify their quest for social advancement by claiming that their inherent qualities made them predisposed to acquire wealth and power.

Besides practical ideas for self-improvement, Combe offers another line of defense to cushion the blow of losing one's free will: evolution. "The constitution of this world," he observes, "appears to be arranged in all its departments on the principle of gradual and progressive improvement. Physical nature itself has undergone many revolutions, and apparently has constantly

106   WRITING THE BRAIN

advanced" (Combe 1841, 4). Humanity, he lets his readers know, can speed up this slow improvement by adhering to the moral and political lessons of phrenology. Infused with quasi-Lamarckian ideas on the inheritance of traits, Combe argues that the behavior and experiences of the parents, especially of the mother, are passed down to the children.

This he claimed to be the case not only for mentally questionable behavior but even for "impressions on the mind of the mother," which can influence the child's mind in utero (Combe 1841, 176). As an example of the latter, Combe describes the case of a shoemaker whose son was "in a state of idiocy" even though the mother was of "sound mind" (1841, 176). Combe's diagnosis is startling: The mother had passed a visual impression on to the child. The father, according to Combe, recalled her meeting an "idiot" while pregnant. "That idiot made a strong impression on his wife; that she complained that she could not get his appearance removed from her mind . . . [consequently] his son was weak in body from birth, and silly in mind, and had the slouched and sloven appearance of the idiot" (1841, 176–77). Self-improvement is thus a multigenerational task. Such an evolutionary perspective allows Combe to transcend at least some of the dreaded material determinism that attached itself to early phrenology. Self-reform might not "fix" an individual's problematic character traits, but it can affect the next generation. While Combe's message is thus couched in a language of optimism, it packs just enough threat to the coming generation(s) to establish itself as an essential intervention. Truly staying sane suddenly required constant self-policing.

In Combe's psychology, people either use their propensities wisely or abuse them, accordingly reaping rewards or punishments that often are passed down through generations. This outlook on life certainly holds narrative potential: crimes of fathers to be read in their children's features; a sick body disclosing moral decrepitude; a moral hero relying on his peculiar talents to achieve his goal; a villain abusing his abilities for vice and getting his just comeuppance. The vision of Combe's theory is that of a self-regulatory society, and literary contrivances like coincidences, happy endings, or multigenerational guilt now had a scientific theory to justify themselves. What Richard Stang has termed "mid-Victorian realism" thus *did* have its own "pseudo-scientific connections" (1959, 5)—even though Stang might have overlooked, from his 1950s perspective, this peculiar set of theories.

Phrenological Victorianism

The works of Charles Dickens—the literary master of many a nineteenth-century pseudoscience and a particular darling of the phrenologists[11]—are especially brimming with the Realism that phrenology had to offer. From his early works to his last, Dickens creates plots by summoning up a complex set of characters, made up of interlocking systems of propensities, that struggle through challenges created by out-of-placeness or incompatible psychological makeups toward the ultimate goal of restoring psychosocial homeostasis. Andrew Sanders has described Dickens's political vision as driven by an "idea of alienation [that] is centred ... on the experimental notions of dislocation, disjunction and disjection" (1994, 22). His vision of conflict is one not of class but of the very idea of *psychological* belonging promoted by phrenology. Although some of Dickens's fascination with the science had waned following a skull reading by the American Fowler brothers, which he experienced as invasive (Stolte 2022, 74–77), phrenology is still the best framework for analyzing Dickens's interlocking literary minds on a societal, meso level.

While the aforementioned key themes of phrenological narratives can be found in many of Dickens's works, they are perhaps nowhere as pronounced as in his penultimate novel, *Our Mutual Friend* (1864–65). Not only does it feature several phrenological puns[12]—like an odd "gentleman with [a] *lumpy* forehead" disrupting a discussion about the "*Constitution* of Britain" by exclaiming "ESKER" (meaning "ridge") (Dickens 1971, 178)—as well as examples of behavior phrenologically transmuted into diseases of the next generation,[13] but it at times reads like a lesson in phrenological Realism. As such, the novel might appear a "mere bundle of eccentricities, animated by

[11] Dickens, in a letter to Charles Lever, stated about phrenology that he "believe[s] in it, in the main and broadly" (1933, 13), and many of his works are littered with puns about and references to its theories and key symbols. His embrace of the science was returned by phrenologists, who often discussed his works in detail in their scientific journals. Upon Dickens's death, a eulogy in the *Illustrated Annuals of Phrenology and Physiognomy* particularly praised the author's "large brain" and his pronounced "intellectual lobe" (*Illustrated Annuals* 1873, 45).

[12] These puns have been used to suggest that Dickens kept an ironic distance from the science (e.g., Marlow 1993, 228; Mengham 2001, 4). Juliet McMaster rightly objects, arguing that regarding phrenology, "it is not that Dickens scorns the reading [of features of the head] as signs of character: he objects to the *misreading* of them" (1987, 21).

[13] One example is the alcoholism of Jenny Wren's father, echoed in his daughter's physical disability. The book describes her as "a dwarf... a something" with "queer" legs and a "bad" back (Dickens 1971, 272). Phrenologists, in line with Combe's progressive theory of soft inheritance, generally understood dwarfism, physical deformities, or even gout as expressions of the vices of "intemperate or scrofulous" parents (Kneeland 1851, 351). On Wren, see also Boehm 2013, esp. 151–53.

108 WRITING THE BRAIN

no principle of nature" to those not subscribing to the (pseudo)science—as, in this case, Williams James's brother Henry (in Stolte 2022, 4).

As in the majority of Dickens's works, actual social rising in *Our Mutual Friend* is a recipe for ridicule or disaster. The novel begins and ends in stasis: Wealth is to be passed from father to deserving son, made proper by an arranged marriage. Disbelieving of his father's wisdom, said son fakes his own death to assess whether the moral fortitude of the people affected by this inheritance lines up with their new lot in life. Through a series of twists and turns, the son ultimately finds confirmation: The servants are indeed natural servants and the wife a natural companion. "Misfortunes," Combe states in his *Constitution of Man*, "can be traced to the error of having placed human beings, decidedly deficient in moral or intellectual qualities, in situations which demand these in a higher degree than they possessed them" (1841, 200)—and these very "misfortunes" form the core of the novel, from its overarching plot(s) down to even its side characters.

Our Mutual Friend features a number of protagonists that seem almost to have been created with Combe's book in front of the author. Even a minor figure like the swindler Alfred Lammle radiates this particular brand of proto-psychology. While himself a crook and unable to achieve his dreams of becoming rich, he nonetheless serves as a vessel of justice, of sorts, in the story by ensuring an even less reputable character's just comeuppance. In a rare moment of insight, we find Lammle reflecting on his own psyche when faced with the naive servant (and temporary inheritor) Boffin being mistreated. Lammle informs his wife and fellow swindler:

> [People undeserving of Boffin] rouse my indignation, my dear, against the unworthy persons, and give me a *combative* desire to stand between Mr Boffin and all such persons. Why? Because in my lower nature I am *more worldly and less delicate*. Not being so magnanimous as Mr Boffin, I feel his injuries more than he does himself, and feel more capable of opposing his injurers. (Dickens 1971, 708)

From a phrenological perspective, Lammle has an overdeveloped organ of Combativeness (see Figure 12, no. 6), considered by Combe one of the "lower organs" (1841, 301). It was believed to be situated on the back of the head and shared with "cocks" and "dogs" (Combe 1841, 288). "Combativeness" was conceptualized as a driving force for self-preservation, granting the "courage to meet danger and overcome difficulties [and a] tendency to

SKULLS AND SOCIETY 109

oppose ... unjust encroachments" (Combe 1841, 53)—the main premise of phrenology, of course, being that no organ is bad by itself. Prone to fights and verbal aggression, Lammle consequently does so out of an inflated sense of combating injustice. Justice, in phrenology, is not a unitary concept but the result of negotiations between different and often opposing mental systems, in particular Combativeness, Benevolence, Veneration, and Contentiousness. With the more "delicate" higher organs underdeveloped in Lammle, as he admits himself, his survival instincts tend to be petty and merely self-interested. Only in the rare case when self-interest and justice align can it be employed for its best function.

Putting himself in opposition to Mr. Boffin, Lammle also discloses the very lack of Combativeness that turns said servant into a perfect example of out-of-placeness. The most loyal of the deceased father's servants, he (temporarily) inherits all of his wealth, while the son (believed to be dead) goes undercover as Boffin's new secretary. Boffin and wife then become this tale's parody of social risers, their clumsy inability to dress and behave in line with their new standing charming the reader while underscoring their inherent subservience. What makes for a good servant rarely makes for a good gentleman—and Combe supplies quite practical hints at picking the right traits, based on their cranium.

Knowing his audience, Combe, throughout his various works, dedicates a substantial number of pages to spotting the perfect servant. Before detailing specific cranial considerations for hiring the best candidate, he summarizes:

> The best combination of the intellectual organs for a servant, is that which occurs when the lower region of the forehead is large, the middle region immediately above the nose, up to the line of hair, is also large, and the upper lateral region full. (Combe 1860, 416)

With his hanging cheeks and large forehead (Dickens 1971, 90), Boffin not only looks the part but backs it up with the perfect qualities behind his bumps and ridges. The dangers of hiring problematic servants were known to the heir of *Our Mutual Friend* as well as to Combe; both were quite aware of the much-discussed "servant problem" of the day, with "uppity and unruly" behavior believed to be "aris[ing] from ... vicious character flaws" of domestic staff (McCuskey 1997). A "master," Combe echoes the worries of our mysterious heir, "must employ individuals who know the moral law, and who

110 WRITING THE BRAIN

possess the desire to act under it; otherwise, as a punishment for neglecting this requisite, he may be robbed, cheated, or murdered" (1841, 238).

Combe's "moral law" is his attempt to formulate a "science" of morality (cf. Combe 1841, 232), based on the idea that higher moral sentiments have to guide and channel the lower "animal propensities" into actions society deems right and proper.[14] There thus exists a dualism within each individual that is mirrored in society on a macro or meso level: Higher sentiments correct lower sentiments, and lower sentiments fulfill duties left vacant by higher sentiments. The servant-master relationship is thus biologized as the external mirror of a complementary duality that exists within each person: Whereas Boffin's highly developed Wonder, Veneration, Love of Approbation, Imitation, and Philoprogenitiveness make him a trusting, caring servant who responds well to praise, his relative lack of Combativeness, Acquisitiveness and Cautiousness renders him a rather poor bourgeois landowner.

John Harmon, the heir who faked his own death and the novel's eponymous "mutual friend," on the other hand, seems like Boffin's mirror image: overcautious and distrusting to an almost paranoid degree, prone to skepticism, and described as reserved, awkward, and somewhat cold. "The natural qualities of the master or mistress to those of the servant," Combe cautions, "must be attended to" and have to correspond (1860, 417). Here, he formulates a somewhat more conciliatory response to the "servant problem" (it may not *just* be the servant's fault, but a mismatch in propensities on both sides), while also cementing social stratification as a question of brain development. In Boffin and Hammond, we find such a complementary relationship: a naturally just and ultimately wise master, whose rougher edges must be softened by a caring, loving servant, content with his lot.

Where the plot of the complementary pair of Boffin and Hammond is resolved by reversing their "unnatural" positions in society back to their proper place, another set of characters poses a more complex phrenological issue: love. In another major plotline, we find a different antagonistic couple: the gentleman Eugene Wrayburn and social riser Bradley Headstone, competing for the affections of Lizzie Hexam. In their pursuit, they offer a phrenological lesson on the conflict between the so-called Energetical Faculties (Firmness, Self-Esteem, Approbativeness, Destructiveness,

---

[14] Fascinatingly, Combe approaches moral relativism here, suggesting that different societies define different qualities as proper or right, before ultimately preempting Victorian outrage by assuring his readers that indeed such societies have yet to catch up to Western society, described as more advanced in a progressive view of (non-Darwinian) evolution.

Cautiousness) and the Intellectual Faculties (Causality, Comparison, Human Nature, Constructiveness).

Bradley Headstone—yet another Dickensian example of upward mobility conflicting with inherent mental makeup—clearly exhibits an overdevelopment of the former qualities, while lacking the latter to keep his lower propensities in check. Described as "proud, moody, and sullen" (1971, 267), Headstone is haunted by the shortcomings of his brain, as Dickens's phrenological reading of him suggests:

> There was a kind of settled trouble in the face. It was the face belonging to a naturally slow or inattentive intellect that had toiled hard to get what it had won, and that had to hold it now that it was gotten. He always seemed to be uneasy lest anything should be missing from his mental warehouse, and taking stock to assure himself. (1971, 267)

Headstone exhibits overactive, unchecked Self-Esteem in its phrenological sense (not to be confused with its more modern sense as positive self-image):[15] "an intense feeling of egotism" and an overinflated view of one's "own respectability" that lead to "censoriousness and envy" (Combe 1860, 235). Having raised himself out of poverty and achieved a moderately respectable standing as a schoolmaster, he feels entitled to the love and admiration of the noble pauper Hexam and lacks the intellectual ability to handle her rejection. "An Individual," Combe chimes in, "possessing an active temperament, and Self-Esteem, Combativeness and Destructiveness, larger than [the Intellectual Faculties] will be impatient of opposition and contradiction" (1860, 541). The dangers are obvious: "An abuse of Destructiveness leads to murder" (Combe 1841, 367). And that is exactly what the out-of-place Headstone attempts.

The target of his wrath and Hexam's love, Eugene Wrayburn, at first seems an unlikely hero to this villain. While he is certainly the polar opposite of Headstone, this fact makes him, from a phrenological perspective, only lacking in other features. While he is patient, his underdeveloped Firmness renders him aimless and utterly without drive; while he is clever, his lack of Approbativeness makes him a lousy lawyer; and while he is witty, his want of Cautiousness almost costs him his life. The conflict of Intellect

---

[15] While phrenology did not coin the term, it certainly popularized it and laid the terminological groundwork for modern (pop)psychological concepts of self-esteem.

112 WRITING THE BRAIN

against Energy, Wrayburn against Headstone, comes to a head on a moonlit riverbank:

> In an instant, with a dreadful crash, the reflected night turned crooked, flames shot jaggedly across the air, and the moon and stars came bursting from the sky.
>
> Was he struck by lightning? With some incoherent half-formed thought to that effect, he turned under the blows that were blinding him and mashing his life, and closed with a murderer, whom he caught by a red neckerchief—unless the raining down of his own blood gave it that hue.
>
> Eugene was light, active, and expert; but his arms were broken, or he was paralysed, and could do no more than hang on to the man, with his head swung back, so that he could see nothing but the heaving sky. After dragging at the assailant, he fell on the bank with him, and then there was another great crash, and then a splash, and all was done. (Dickens 1971, 766)

Wrayburn turns during the attack and his head swings back, suggesting at least one severe blow to the front of his head, where phrenologists believed the Intellectual Faculties and language resided (Combe 1860, 454–55).

Shortly afterward, we find Wrayburn, who miraculously survived the attack, mute and partially paralyzed, his "disfigured forehead" (Dickens 1971, 769) now in the loving care of Lizzie Hexam. As a result of his injury, Wrayburn changes—once a prime example of idleness and ennui, he now discovers determination, energy, and even passion within himself, leading him, for instance, to openly cry for the first time. Headstone, of course, punishes himself by committing suicide.

The fact that his attack took place by a river has led many critics to address his change of character in religious terms: Wrayburn is "being born again and beginning a new life" (Page 1984, 228). Phrenology opens up a reading that is much more direct than a metaphorical holy baptism in the notoriously sewage-infected Thames. Even the names of the two men point to the common trope of head injury as character change: "head stone," "wray burn"—a stone to the head reveals[16] fire within. The "flames [shooting] jaggedly" (Dickens 1971, 766) in front of his mind's eye already hint at Wrayburn's transformation from Intellect to Energy. His overactive Intellectual Faculties temporarily deactivated or at least subdued by severe

---

[16] "Wray," of course, meaning "reveal" or "disclose" in the Middle English of Chaucer.

SKULLS AND SOCIETY   113

cranial damage, the once muted qualities from the posterior of his skull—
the Firmness and Combativeness to pursue a goal with intensity, but also the
Amativeness and Affection to reciprocate Hexam's feeling in kind—finally
come to influence his behavior for the better.

In a worldview where the biological makeup that shapes one's skull from
within largely determines one's character and behavior, true change is an
extraordinary event that necessitates a traumatic, outside force. Dickens's
goal—and the overarching goal of phrenological Realism—is to argue for
restoring proper balance between opposing, complementary poles on the
societal and individual levels. The sweeping changes threatening the social
order of Victorian society can thus be looked at as temporary, requiring fundamental adjustments only where innate abilities conflict with social position. Certainly, Combe and Dickens agree, it takes quite a different cranium
to rule, to own, or to administer than it does to serve and venerate. There
is thus a proper, almost inevitable, structure to society, dictated by biology,
that makes room for benevolence toward the poor while literalizing said
Benevolence as an innate ability of the ruling class. Benevolence is not a temporary alleviation of an underlying problem but a reflection of natural order.

Phrenological Americanism

Such a class-based view, however, certainly should have raised alarms for
one of Dickens's most ardent admirers abroad: Walt Whitman—the self-
identified radical democrat and socialist by association. His infatuation
with phrenology can be traced to his earliest writings[17] and plays a key role
in his ever-developing engagement with the nation in *Leaves of Grass*, one
edition (1856) of which was published by a phrenological press. His poetical use of phrenological concepts to give voice to same-sex attraction has
received ample scholarly attention (e.g., Reynolds 1995, 247–50). Still, even
as an old man, when phrenology could no longer be considered a valid science, Whitman recalled the compliments the theory had offered about his
skull (in Traubel 1906, 282). His phrenological reading of 1849 by Lorenzo

[17] In his essayistic newspaper series "Letters from a Travelling Bachelor" from 1849, six years prior
to the first edition of *Leaves*, Whitman already uses phrenological vocabulary in a fairly detailed
manner to describe characters: "He is a fat, stolid looking old man. Benevolence and philoprogenitiveness have made themselves scarce on his skull-cap," he writes at one point. "Neither a physiognomist nor a phrenologist would have been pleased with her face and head," he observes at another
(1849).

114   WRITING THE BRAIN

N. Fowler (1811–1896)[18] became such an important document for the poet that he referenced it not only in anonymous self-reviews of *Leaves*[19] but also in his prose works: A character (a Spanish dancing girl, no less) in his Dickens-inspired novella *Life and Adventures of Jack Engle* is described using expressions lifted from his 1849 assessment.[20]

The earliest accounts by Whitman on the topic of phrenology likely stem from his time as the editor of the *Brooklyn Daily Eagle* in the 1840s. We can find not only a glowing review of Spurzheim's works on phrenology that was likely authored by the future poet[21] but also a review of lectures on phrenology that sheds light on how Whitman initially encountered the science. One of the first pieces the author wrote for the *Daily Eagle* (it came out early in his first month as its editor, in March 1846), it gives a brief account of the Brooklyn lectures of Lorenzo N. Fowler, the poet's later co-publisher:

> The tri-weekly audiences of Professor Fowler [that were] generally large at first, when the admission was gratis, have fallen off in number since the sixpenny exaction was instituted. . . . Phrenology, we are taught, is the only thing needful to unfold all the mysteries which beset this life, and teach us the mode of perfecting ourselves for eternity. . . . We hope the citizens of Brooklyn will be aroused to a proper interest in this great subject of the human mind as interpreted by the phrenology of Professor Fowler.—The examination of heads which usually terminates the lecture is immeasurably funny. After picturing out the character of the subject, he has a curious way of asking him or her whether he has not been correct . . . as if any one practised the great dictate: "Know thyself." ("Lectures" 1846)

[18] "Size of head large, 23 inches. Leading traits appear to be Friendship, Sympathy, Sublimity, and Self-Esteem, and markedly among his combinations the dangerous faults of Indolence, a tendency to the pleasures of Voluptuousness and Alimentiveness, and a certain reckless swing of animal will" (Whitman 1856, 362).

[19] In his self-review in the *Brooklyn Daily Times* of 1855, Whitman provides a detailed footnote of his phrenological reading (see previous note). The review is reprinted in the annex to the 1856 edition of (1856, 362). Whitman also authored an anonymous self-review for the *American Phrenological Journal* in the same year that references the science.

[20] See Schöberlein and Blalock 2020. Indeed, there seems to be a triangulation between Fowler, Whitman, and Dickens: Some of the ideas about "proper maturation," as Jason Stacy has pointed out to me, that Whitman takes from the Fowlers, the Fowlers likely lifted from Dickens's *Dombey and Son*.

[21] The short review praises phrenology for having withstood the "waves of detraction" it had to face. Now, the piece observes, not only has phrenology "gained a position, and a firm one, among the sciences," but one might even say its proponents "have gained a victory" ("Notices of New Books," 1846). For more on the attribution of this piece, see Brasher 1958.

SKULLS AND SOCIETY 115

Even while the tone of his review, and that of a follow-up piece two days later (which lambasted the whole performance as "pretension and absurdity"), is somewhat critical of Fowler's teachings, we might want to stop short of Jerome Loving's assertion that Whitman "thought the system an intellectual fraud" (1999, 104). While, if indeed authored by Whitman,[22] the follow-up review indicates that it took the future poet some weeks to warm up to the person of Fowler, one can hardly overlook the fact that it reveals these lectures to have been a go-to activity for the young reviewer: He attended so many times that he could comment on shifting audience numbers as a result of the lecturer beginning to charge an entry fee (a fee that Whitman was, apparently, very willing to pay, even after knowing quite well what Fowler had to offer). Even later in life, Whitman seemed to recall the lectures of March 1846 and did so fondly. In a conversation with Horace Traubel in 1890, he not only mentions coming "across an old phrenological chart from L. N. Fowler . . . given before 'Leaves of Grass'" but also echoes the wording of the 1846 review: "The science, or what-not, called by many names—was always *funny* to me—using the word *funny* in the sense that includes power, mystery—but I recognized its value" (in Traubel 1992, 109, emphasis his).

Lorenzo Fowler was hardly the "professor" Whitman made him out to be. Like Combe, upon whose teachings he based his works, Fowler was a convert to the field of psychology: He had studied to become a minister, alongside Henry Ward Beecher, but then switched to the more exciting (and lucrative) field of phrenology, offering readings, lectures, and (as a part of Fowler and Wells and *Life Illustrated*)[23] publishing opportunities for like-minded thinkers. While he lacked training in medicine or philosophy of mind, he did

---

[22] Indeed, there is some indication that Whitman may not have written this follow-up review. A stylometric authorship assessment failed to confirm or even suggest Whitman as the author of the March 7 piece (the assessment employed the corpus and method described in McMullen, Stacy, and Schöberlein 2019). Content also sheds some doubt on Whitman's authorship: The personal offense taken by the author of March 7 piece over Fowler's notion that he could read a person's religious leanings on their skull seems at odds with Whitman's rather secular orientation. Given Whitman's love of accents and slang, the negative comments about Fowler's peculiar pronunciation also seem odd. The suggestion of repeated attendance (the main theme of the March 5 piece) is also missing, as is the reference to the "fun" of the performance(s). A tone similar to that of the March 7 review can be found in the newspaper's dismissal of Fowler's lectures as "ridiculous" prior to Whitman's tenure at the paper (February 12, 1846). Less than a year after its scathing March 7 dismissal of Fowler, the paper (then firmly under Whitman's editorial leadership) ran a piece *praising* Fowler, recommending two of his books, and stating, "There can be no harm, but probably much good, in pursuing the study of phrenology" ("Something About Physiology and Phrenology" 1847).

[23] Fowler and Wells also republished a number of Combe's works, including *The Constitution of Man*, as well as Spurzheim's *Reminiscences of Combe*.

116   WRITING THE BRAIN

have one thing to offer to the field: a deeply *American*, democratic take on a thoroughly Victorianized theory.

"To Americanize whatever in science and the arts, is capable of improving or adorning the mind . . . is no less the duty, than it would prove the glory, of every American citizen," exclaims Fowler's first book, co-authored with his brother Orson in 1836 (and a likely blueprint for his 1840s lectures). Since "hitherto, no American work has appeared on this subject [phrenology], stamped with originality of thought," their *Phrenology Proved* now hopes to fill that void (Fowler and Fowler 1836, iii). While largely in line with Combean phrenology and fully embracing its evolutionary notions on the inheritance of acquired traits, the Fowlers' Americanist interventions are telling, especially when considering phrenology's impact on a figure such as Whitman. The most significant twist on Combe that they offer is their interpretation of a specific organ that Whitman would come to see as one of his own "leading traits": Self-Esteem.

A somewhat obscure, Miltonian term[24] until midcentury—Webster's dictionary began listing it only in 1848—we have American phrenology to thank for popularizing Self-Esteem as a psychological concept (cf. Harper n.d., s.v. "self-esteem"). Before its revamping in the United States, the term was used synonymously with "self-love" or "undue pride" and was understood as something that constantly needed to be kept in check. Gall and Spurzheim's original definition was, perhaps unsurprisingly, almost entirely negative:

> Gall first found this organ in a beggar. . . . Proud persons and those who, alienated by pride, imagine themselves to be emperors, kings, ministers, generals &c. possess it in a high degree. This faculty gives us a great opinion of our own person—self-love or self-esteem. . . . There are a greater number of mad men than of mad women, who are alienated by pride. (Gall and Spurzheim 1815, 156)

Instead of Self-Esteem enabling a secure sense of identity, the science of Gall and Spurzheim originally conceptualized it as a constant *threat* to a proper self, even leading to self-alienation. It was thus not considered an Intellectual

---

[24] Why the term would have appealed to early phrenologists becomes clear in its context in *Paradise Lost*, where it functions both as an acknowledgment of one's own psychological consistency over time while also necessitating balancing and management: "self esteem, grounded on just and right / Well manag'd that skill the more thou know'st" (Milton 2012, 356).

Faculty but was one of the "sentiments common to Man and the lower Animals" (Combe 1860, 231).

Certainly, Whitman would not have bragged about his pronounced organ of Self-Esteem in his anonymous self-reviews had that definition continued to dominate in American phrenology. But in its place, the Fowlers offered a much more agreeable take on Self-Esteem—a definitional twist that constitutes nothing less than a reorientation of the entire science toward it:

> The proper office of this faculty is to create, in the bosom of its possessor, a good opinion of himself; of his own character and opinions, and of whatever belongs to, or proceeds from, himself; to beget an esteem and respect for himself; to feel satisfied with himself . . . ; to give a manly tone to the character and turn to the conduct, and a dignified, erect attitude and bearing to the person, and thus, to exert an important influence in elevating and ennobling the character of man. And what is still more important, it gives that innate love of personal liberty and independence, and of religious freedom, so deeply seated in the nature of man. (1836, 113–14)

Not shy about their political leanings, the Fowler brothers then launch into a full-throated defense of the organ that would certainly have appealed to the political radical Whitman:

> The existence of [the faculty of Self-Esteem] demonstrates the position, that the feeling or principle of liberty and of equal rights is inalienable, and inherent in the very nature and constitution of man; that, therefore, it can no more be destroyed than hunger or love; that a purely republican and democratick form of government is the only one adapted to the nature of man and the *only* one calculated to secure universal satisfaction and happiness; and that the subjugation of man by his fellow-man,[25] is an open violation of the principles of human nature. . . . [S]hould things progress, for sixty years to come, as they have done since the Revolution, this nation, the birthplace and the cradle of liberty, will be ruled by an aristocracy, not

---

[25] While remaining silent on the topic of American slavery, the Fowlers clearly demonstrate their sentiments in their passages about Self-Esteem: "There is no danger that this feeling will ever be extinguished; but, in case the subjugation and servitude of man, in any form, should be carried to a very great length, there is danger, ay, a moral *certainty* of a *revolution*, and a revolution, too, attended with a violence proportionate to the pressure laid upon it" (1836, 118)—especially considering "the African race as found in America . . . possess[es] in general either large, or very large . . . self-esteem" (1836, 31).

118   WRITING THE BRAIN

of government, but of monopoly, of wealth, &c. far more tyrannical than any nation under heaven. But . . . man's nature is unalterable; the spirit of Seventy-six, and the love of liberty, *will live* and will *increase* (1836, 118–19).

The Fowlers discovered a truly Jeffersonian faculty, turning what was once undue pride and self-alienation into the physical locus (and in turn proof of) American-style individualism, manliness, and democracy.

Masculinity, self-contentment and self-celebration, a sense that mind and democracy correspond—there's much Whitman would have appreciated in the works of the Fowlers. And indeed, we can see clear traces of their words in *Leaves of Grass*, where "self-esteem"[26] is repeatedly referenced as a core component to Americanism. The 1855 prose preface, for instance, states that "the genius of the United States is . . . best or most in its common people," providing "their *self-esteem* and wonderful sympathy" as evidence of it, and claiming that the American dialect "is the chosen tongue to express growth faith *self-esteem* freedom justice equality friendliness amplitude prudence decision and courage" (Whitman 1855, iii, xii). At the same time, the preface claims the "phrenologist" as one of the "lawgivers of poets," raising them to the same level as the "anatomist chemist astronomer geologist . . . spiritualist mathematician historian and lexicographer" (Whitman 1855, vii).

The poet's actual responses to the phrenologist as a lawgiver can be observed in Whitman's "Poem of Many in One" (its title oscillating with the possibility of being a phrenological pun). It first appeared in the 1856 edition of *Leaves of Grass*, published by Fowler and Wells, and ultimately retitled "By Blue Ontario's Shore."[27] Drawing heavily from the prose preface, the poem is Whitman's call for a republican, democratic mindset (and for a poet to bring it about). While the connection to Emerson's "The Poet" appears quite obvious, the piece also engages the politico-phrenological thought of the Fowlers: "Have you learned the physiology, phrenology, politics, geography,

---

[26] Throughout his literary and private life, Whitman's use of "self-esteem" always also radiates its phrenological meaning, at times disclosed by his capitalization of the term (as is typical in phreno-logical textbooks). As late as 1888, Whitman uses craniological cartography as a handy metaphor with regard to this organ: "Eighty millions of Tartars may shave the top of the head from Comparison to Self-Esteem" (in Traubel 1906, 36)—that is, from the middle of the forehead to the upper back of the head, somewhat similar to a tonsure.

[27] The Fowlers' hometown, Cohocton, is located fifty miles south of Lake Ontario. Should the eponymous trip to its lakeshore have been based on a real-life event, Whitman may have passed through it, perhaps even triggered by his personal acquaintance with the Fowlers in the late 1840s and early 1850s.

SKULLS AND SOCIETY 119

pride, freedom, friendship, of my land?"[28] (Whitman 1856, 192), the voice of the piece inquires of the poet.

The "phrenology . . . of [the] land" we can find everywhere in the oeuvre of the Fowlers, but perhaps nowhere as pronounced as the articles of 1849, the year Lorenzo Fowler measured Whitman's head. In particular, a piece in the Fowler-edited *American Phrenological Journal and Miscellany* on "Republicanism" seems to anticipate Whitman's "Poem of Many in One." Where Whitman would ultimately come to ask himself, "Do you see who have left all feudal processes and poems behind them, and assumed the poems and processes of Democracy?"[29] (1892, 271), Fowler and his brother inquire (anticipating, or perhaps providing, the title of Whitman's poem): "Do you discard all forms of aristocracy, and adopt the republican principle of E PLURIBUS UNUM—many in one?" (Fowler and Fowler 1849, 215).

Both pieces, "Republicanism" and "Poem of Many in One"/"By Blue Ontario's Shore," understand the Declaration of Independence as a foundational, *organic* event that constituted a releasing of pent-up, inherent qualities of the mind (especially Self-Esteem) while also posing an evolutionary challenge for the future. Turning to himself, the poet wonders:

> Have you consider'd the organic compact of the first day of the first year of Independence, sign'd by the Commissioners, ratified by the States, and read by Washington at the head of the army?
> Have you possess'd yourself of the Federal Constitution? (Whitman 1892, 271)

Underscoring quite aptly the shared language of republicanism and phrenology—"head," "Constitution," even "Independence"—Whitman subscribes to the notion that the Declaration was foremost a *mental event* that triggered a societal change that was still ongoing. "Its progress," the Fowlers write of the Declaration of Independence, "was at first slow. . . . It released the American pockets from the grasp of despotism; but it still left the American MIND, religion, framework of society, and feeling, as thoroughly aristocratic as the

---

[28] Here phrenology not only is listed but almost seems to structure the list: It is a physiological "geography" of the skull that discloses "pride, freedom, friendship" (phrenological buzzwords) and has a political program.

[29] In 1856, Whitman is still more vague, asking about "described processes and poems"; the clarification to "feudal" comes in 1871.

120   WRITING THE BRAIN

English mind itself." But now, an "awakened . . . spirit of human reform" sweeps the country (Fowler and Fowler 1849, 208), and Whitman agrees:

> Already a nonchalant breed, silently emerging, appears on the streets,
> People's lips salute only doers, lovers, satisfiers, positive knowers,
> There will shortly be no more priests, I say their work is done,
> . . .
> Justice, health, *self-esteem*, clear the way with irresistible power;
> How dare you place any thing before a man? (1892, 272)

"The land of mental society," the Fowlers state, "must be put under the highest possible state of cultivation" (1849, 209). Like Walt Whitman, Lorenzo and Orson Fowler argue that republicanism and aristocracy are *cognitive* qualities above all and come to the forefront in everyday life, as they illustrate using one of Whitman's most beloved tropes—a meeting on the street:

> The distinctive principle of this great association [of the United States] is the Union of the Many for the Good of All. . . . Every one will no doubt . . . acknowledge the truth of this principle; yet the first rich man you meet, ten chances to one but you will experience a kind of reverence for him, as if he were something more than a human being; reverence him, not as a man, but as a RICH man. You thereby practically acknowledge your own aristocracy (1849, 214)

Whitman's nonchalance reflects the Fowlers' republican mindset: It acknowledges only actions and behavior, not status—it sees only "man" and refuses to put any epistemological qualifier, "any thing" like the word "rich," in front of it.

The "highest cultivation" of republicanism is thus a call for self-reform. The Fowlers and Whitman, both putting out self-help manuals in the 1850s,[30] call for a bottom-up, democratic renewal:

> Have you thought there could be but a single Supreme?
> There can be any number of Supremes—one does not countervail another any more than one eye-sight countervails another, or one life
>   countervails another.

---

[30] For instance, Whitman's "Manly Health and Training" of 1858.

SKULLS AND SOCIETY 121

All is eligible to all,
All is for individuals—all is for you. (Whitman 1856, 180–81)

In its final version, the poem is even more direct, claiming that the "whole theory of the universe is directed unerringly to one single individual—namely to You" (Whitman 1892, 273). If Whitman learned anything from Fowler, it is that the promise of America's future lies in the mind of each individual and radiates outward from there.

Even when castigating, among other things, zealotry,[31] monopolies, and the subjugation of women, their call to "adopt, both in feeling and practice, the thought of our governmental institutions—BROTHERLY LOVE AND EQUALITY" (Fowler and Fowler 1849, 215) is still a reformist vision, not a revolutionary one. This becomes especially clear in their stance on slavery. While sharing a condemnation of the institution, Whitman and the Fowlers seem reluctant to want to overthrow it outright. "I swear I dare not shirk any part of myself," the bard-turned-nation of "Poem of Many in One" proclaims, and finds the country's "crimes, lies, thefts, defections, slavery" contained within "you and me," while embracing both the "tremulous spreading of hands to shelter [slavery]" and "the stern opposition to it" (Whitman 1856, 199, 187). The Fowlers elaborate:

Lay down the sledge-hammer and the broad-axe of denunciation. Let anti-slavery men pour out no more vials of bitter wrath against slavery, as such; . . . In short, let the machinery in progress to arrest these evils be laid aside, and this one instrumentality be substituted, namely, the diffusion of republican principles. Let every one begin with himself, and see wherein and how far he is republican, and make himself wholly democratic. . . . In short, let him bind himself to every member of the human family; for republicanism is nothing less than the universal love of man for man. (1849, 213)

The "body politic" (Fowler and Fowler 1849, 215) of "Many in One" begins and ends on the level of the single mind. The rest will surely follow suit. European phrenology envisioned self and society as a compensatory,

---

[31] Like Whitman, Fowler and Fowler call for a fading away of priests: "Perhaps you are religio-aristocratic, and pay not only respect to your clergymen, but are led to believe and do what you should not believe or do. Call you this republicanism? I call it aristocracy, and in its very worst form" (1849, 214). In the true republican society, for the Fowlers, nobody but oneself ought to control one's beliefs or actions.

122   WRITING THE BRAIN

self-regulatory web of connections, with a lack in one mind corresponding to overdevelopment in another. The Fowlers, on the other hand, celebrate the individual psyche as potentially perfect in itself, with its very structure promising a future of equality and justice (instead of confirming one's proper place in the social hierarchy).

What began as a determinist vision in Gall and Spurzheim and aged into a theory of the human psyche as embattled (and thus in constant need of checking and balancing) in Combe became a utopian vision for self-reform in the work of American phrenologists like the Fowlers. The Combean need for self-control within society morphed into a call for self-betterment that in turn reshapes society: "It is the fundamental principle of Phrenology," the Fowlers claim, "that EVERY FACULTY IS ORIGINALLY GOOD, and its PROPER EXERCISE, VIRTUOUS." It thus follows that "all vice, and all sin, are the *excess* or *perversion* of *some good faculty*" (Fowler 1836, 45–46). Gall and Spurzheim saw a proper purpose in each faculty of the brain, when developed normally, but denied that either is good or bad by itself. This neutral stance, with a subtle twist of perspective, turns into a promise of perfectibility in the Fowlers. Gone is the rigidity of the mind that necessitated blunt-force trauma to experience real change. In its place, we find a brain that, in all its variability from head to head, is perfect and fully capable of attaining utopia, as long as it is willing to embrace its nature. Looking at his phrenological organs, Whitman famously agrees: "Welcome is every organ and attribute of me, and of any man hearty and clean, / Not an inch nor a particle of an inch is vile, and none shall be less familiar than the rest" (1855, 14).

Whether Whitman happened to read the "Republicanism" essay or gleaned its contents from the Fowlers' various essays, books, or lectures, their influence on *Leaves* is hard to overlook. The formative years of his book fall squarely into the height of Whitman's fascination with the science of the Fowlers and shares not only their terminology but also their politico-philosophical outlook. And it wasn't a short-lived fad: Scholars have traced phrenology into Whitman's ideas about writing epitaphs (Mackey 2005, 21–39), his politics of male-male affection in the 1860 edition of *Leaves* (Mullins 2006, 171), and even his theories on child-rearing (Aspiz 1966).[32] Still, there

---

[32] On Whitman and phrenology, see also Borst 2014; Burrell 2004, ch. 11; Dalton 2002; and Wrobel 1974. Especially Wrobel's piece is fascinating in its emphasis on similarities between transcendentalism and phrenology, while still presupposing that the former must have come before the latter, and that the latter takes on a subservient position at best and is discarded or at least muted in Whitman's later years. The politics of phrenology remain (even in regard to Whitman) sorely underdiscussed,

SKULLS AND SOCIETY 123

is a scholarly tendency to merely see these moments as some "sprinkled phrenological terms" (Reynolds 1995, 247) or as reducing his engagement with the science to a "borrow[ing] of two terms . . . [albeit] shaped and revised" (Mullins 2006, 171). What, one has to wonder, would our reading of *Leaves of Grass* look like if we encountered it *first* through the gaze of phrenology as its "lawgiver"—not, say, via Emersonian transcendentalism (an influence backed up by significantly less textual and biographical evidence)?[33] Given that the young Whitman tended to consume his high literature mostly in the form of bawdy theater adaptations, his philosophy in the form of newspaper pieces and summaries, his music via comedic operas, and his politics in the audience of rowdy mass meetings, it is perhaps time to reassess how much of Whitman's philosophy is indeed indebted more to the carnivalesque Fowlers than Emerson's heady meditations.

Phrenology's Real

Can these elements in Whitman, Dickens, and other mid-Victorian writers then be considered a "Realism" when much of phrenology feels too counter-intuitive to contemporary conceptualizations of what constitutes the Real? In his discussion of "medical realism," Lawrence Rothfield argues against such a perspective, claiming instead that "neither [in] phrenology, nor physiognomy . . . can one find a systematic vision of this world" (1992, 56–57). Perusing the more philosophical edges of phrenological texts of the day, it becomes clear that this is not the case. The "societal vision" of Combe is one of servants and masters—and the self-help aspect of his work, focused on finding one's natural position in society—whereas the Fowlers are even more up-front about the politics that seep through every page of Combe's works, as they see the promise of America carved into the skull of every human being.

---

aside from phrenologist complicity in racial stereotyping and its role in anti-Native or anti-Black defenses of white supremacy that persist to the present day (Chinoy 2019).

[33] The evidence generally boils down to Whitman sending *Leaves* to Emerson in 1855 and his positive review of an Emerson lecture in 1842. Although praising the lecture as "one of the richest and most beautiful compositions," Whitman in this review struggles to even summarize the lecture's contents, ultimately settling on the claim "it would do the lecturer great injustice to attempt any thing like a sketch of his ideas" (2014, 192). In the end, the review features more detailed descriptions of the dress of women in the audience than of Emerson or his words. When Whitman later claimed that in the 1850s he "was simmering, simmering, simmering; Emerson brought [him] to a boil" (in Reynolds 1995, 82), we can perhaps either hear a clever self-promoter or wonder what brought Whitman to a "simmer" in the first place. For the case for Emerson, see Stovall 1974 and Loving 1982.

124 WRITING THE BRAIN

If the goal of any realism is to achieve an "objective representation of the current social reality" (E. B. Greenwood in Villanueva 1997, 3), one ought to keep in mind, as Piero Raffa has cautioned, that "a true theory of Realism that hopes to escape the limitations of a particular style, cannot or ought not offer any specific definition of reality" (1967, 238).

It is easy to dismiss phrenology as a nineteenth-century curio and a Barnumesque kind of charlatanism and thus to mark the writings of "proper" psychologist William James as the starting point of the psychological "Real" in literature. Still, whether it is a fleeting glance at Whitman's pathos and optimism or a sneer over the "frequently false psychology" of Dickens's characters (Eliot 1856, 33) that triggers such dismissal, one ought to keep in mind that in the "specific definition of reality" these writers subscribed to, their representations could be considered "objective." In place of the still-prevailing metaphysics of the mind, these mid-Victorian writers subscribed to a materialist psychology that had a scientifically justified tool set to shape narrative. Their characters—Dickens's social risers and Whitman's "I" and "You"—are structured around an attempt to represent psychology as it exists not in an idealized, philosophical sense but in everyday life. Still, their characters, major or minor, are rarely made subservient to phrenology as such; their sole purpose is not to prove the theory (thus making it easy to overlook the "systematic vision" of it) but to approximate figures one might meet on the street or hear about from a neighbor. One of the overarching goals of Gall, Combe, and the Fowlers was to make the everyday readable, to disclose "real" patterns behind psyches and behaviors, and to banish randomness by defining the psyche as the constant, complex core of personality.

The psychology of phrenology was explicitly a *social* psychology, arguing that brains and society interact on some fundamental level. The phrenological "I" cannot sustain itself in perfect isolation—even the most Romantically inclined Gallian brain wouldn't last long on a lonely mountaintop or deep in the woods. Phrenology, whether it believed society to radiate outward from the individual or saw the two as always already intermeshed, was the first truly modern psychology of mass society. It is no surprise that the authors most in tune with urban sprawl and big city life embraced it for allowing them to represent the everyday realistically and purposefully. While phrenology's lessons (even in the Fowlers' works) seemed troubling when observed from a rationalist perspective—there was no longer an autarkic, complete self but a mess of over- and underperforming brain systems—they also became a cause for optimism for phrenological Realists. Whatever deficiencies an individual

SKULLS AND SOCIETY 125

brain might disclose, there was always another brain ready to step up and balance it out. And, with a little evolutionary push and self-control, society would almost inevitably progress toward perfection, even when this perfection seemed more of a kind with perfectly oiled cogs in a machine.

This progressive optimism also initiates the problematic politics of literary Realism proper, its embrace of the lower classes coupled with a frequent inability to visualize the need for or achievability of radical change. Nowhere is this more obvious than in Whitman, who later in life found himself constantly courted by radical feminists, socialists, anarchists, and various other breeds of political revolutionaries but who never subscribed to any of their programs outright. In 1888, with his own brain severely damaged by strokes and its deficiencies balanced by the presence of a nurse, housekeeper, and his devoted disciple Horace Traubel, Whitman was asked about these figures. He replied that he had ample sympathy for socialism: "Lots of it—lots—lots. In the large sense, whatever the political process, the social end is bound to be achieved: too much is made of property, here, now, in our noisy, bragging civilization—too little of men" (in Traubel 1906, 221). The crux of phrenological politics à la Fowler is still here. Incomprehensible to the young Traubel, who grew up in a time when phrenology had completely lost its respectability, Whitman still subscribed to the idea of the inevitable progress of republicanism—the prospect of slow self-reform fixing society from the bottom up, the notion of embracing the nature of one's mind as a political program. Whitman's focus is on the constitution of "men," he hints—their inward materiality, not their outward situation. He thus wouldn't subscribe to the "program" of Marxists as such, the old poet interjects, but he is "with them in the result." Sensing the uneasiness of the man digging through his mind and letters, Whitman adds: "Ain't we all socialists, after all?" (in Traubel 1906, 221).

4

Cranial Reconstruction

Racialized Brains and the Psychometric Real

If Mr. Peabody were to donate half a million or so to "reconstruct"
the negro brain—to add, say, some twenty per cent, and diminishing
the cerebellum with a corresponding increase of the cerebrum, in
about that proportion, and thus prepare it for the instruction proper
to white people, there would be infinitely more reason to the proposal to "educate negroes" as there now are.

—From the New York *Day Book*, 1869

The N. Y. Day Book is very sure that the negro is inferior to a white
man, but it is careful not to let the negro have an equal chance. Why?
We suppose it is because the negro, if he had fair play, might, with all
his inferiority, get ahead of the editor of the Day Book.

—From the *Jackson Standard*, Ohio, 1869

Uncommon Minds

After revolutionary preacher and insurrectionist Nat Turner was hanged in
November 1831, following one of the largest uprisings of enslaved people
in the antebellum South, his remains were turned over to local surgeons
for dissection. While the exact afterlives of the specimens made from his
body—especially his skull—remain a mystery (Greenberg 2004, 22–23),
the impulse to violate Turner's body comes as no surprise. Physicians often
played a key role in dehumanizing dead Black bodies directly, by turning
them into morbid mementos, as well as indirectly, by justifying supremacist
racial theories through apparent insights gleaned from dissection. They were
an extension of the gruesome keepsake economy that surrounded lynchings
and judicial killings of Black men and women in the United States until

Writing the Brain. Stefan Schöberlein, Oxford University Press. © Oxford University Press 2023.
DOI: 10.1093/oso/9780197693681.003.0005

CRANIAL RECONSTRUCTION 127

the twentieth century (see Washington 2008, 136), Although who, exactly, carved open the body of Turner and what, if any, scientific interests he or they may have claimed for this purpose are unknown, one focus seemed to have been Turner's head. This obsession with the self-liberated slave's skull stretches well into the twenty-first century (cf. Fornal 2016), underscoring an uneasy, raced interest in "figuring out" the constitutive contents of Turner's cranium.

Indeed, for a society that had, in large part, convinced itself of the happy subservience of its enslaved population, the contents of Turner's head constituted a problem: he was almost universally acknowledged as remarkably smart—without the qualification of "for a slave."[1] His lawyer, Thomas R. Gray, a onetime slaveholder who quickly sold Turner's confession after his death,[2] attests "an uncommon share of intelligence, with a mind capable of attaining any thing" (Turner and Gray 1831, 18) in a tone that T. W. Higginson would later describe as a "sort of bewildered enthusiasm" for the client and his cognitive abilities (1861, 185). Turner's conscious self-fashioning (albeit filtered through his lawyer's writings) as a highly rational, unemotional revolutionary as well as a skilled eschatologist seems to knowingly engage with and disallow casting him as a racial *less than*. As a general marching his troops to battle (on what was supposed to be the Fourth of July) or as a divinely blessed intellect able to calculate the end of days, Turner made sure his mind stood out in what was to be his final stand. Even in describing his killings, which he (via Gray) repeatedly and often with painful detail clarifies to be "murder," Turner claims forethought, planning, and calculation.

Indeed, if we are to believe the words of his lawyer, the term "uncommon intelligence" was Turner's (Turner and Gray 1831, 8). This (self-)perceived superiority, both to his fellow enslaved people and to his white oppressors, Turner saw physically manifested in what he described as "certain marks on [his] head" and body, which Gray tried to dismiss as barely noticeable remnants of "a parcel of excrescences which I believe are not at all uncommon, particularly among negroes" (Turner and Gray 1831, 7). In the heyday of

---

[1] "The intellectual inferiority of the negroes is a common, though most absurd apology," Lydia Maria Child noted two years after Turner's death, "for personal prejudice, and the oppressive inequality of the laws" (1833, 155).

[2] Greenberg rightly cautions that "one must not overestimate the reliability of the stories Gray recorded. After all, Gray was not a disinterested bystander; a one-time slaveholder in Southampton County, he clearly sympathized with the whites who died. . . . Gray may have embellished *The Confessions* to make for more interesting reading. Human error is also likely in a document produced so quickly. . . . For all of these reasons, one must be careful when handling *The Confessions*" (2004, 105).

128 WRITING THE BRAIN

American phrenology, Turner is claiming *extra bumps on his skull*—an exterior mark disclosing an exquisitely developed, large brain within.

For a "race" of people that many phrenologists believed to be characterized by an almost complete absence of the "organ of calculation" and an overdevelopment of the organs of "devotion" and "benevolence" ("Our New Dictionary" 1865, 108–9)—that is, a mind made for servitude—Turner breaks the mold in a troubling way. Even in the racialized depiction by his white lawyer, he acknowledges these stereotypes by subverting them at every point. Turner even states he learned to read so quickly and easily as a child that it predated memory formation.[3] Gray, in his conflicted role of salesperson, lawyer, and assuager of white worries, makes sure to counter such statements right away. Turner's "own account," Gray tells his readers before they hear Turner speak, he believes to contain "a useful lesson, as to the operations of a mind like his, endeavoring to grapple with things beyond its reach." Liberation, Gray reminds his readers, is beyond even Turner's abilities. A Black brain, the lawyer suggests, may at best house a *"dark*, bewildered, and overwrought mind" (Turner and Gray 1831, 4, emphasis mine).

Turner's assertion of superior brainpower and his lawyer's framing were echoed widely—so far, indeed, that Pennsylvania native and physician Robert Montgomery Bird (1806–1854) felt compelled to concoct a racist caricature of the events and its leader to dispel them. Tonally at odds even in his rather morbid 1836 *Sheppard Lee*, its short first-person account of "Nigger Tom" combines anti-Black minstrelsy with rape and graphic violence against women and children—all in the jarring tongue-in-cheek tone of a parody. Inspired by contemporary newspaper accounts of the Turner insurrection (which perhaps even supplied the title of the book).[4] Bird purposely rewrote Turner's insurrection to fit a narrative of Black inferiority. Instead of reading about a single preacher-general, Bird's audience is treated to two lesser brains, resulting in characters that are half-human, racial stereotypes: "Parson Jim," who is "fond of praying and preaching" but quickly turns into a "very active

---

[3] "The manner in which I learned to read and write, not only had great influence on my own mind, as I acquired it with the most perfect ease, so much so, that I have no recollection whatever of learning the alphabet—but to the astonishment of the family, one day, when a book was shewn me to keep me from crying, I began spelling the names of different objects—this was a source of wonder to all in the neighborhood, particularly the blacks" (Turner and Gray 1831, 8).

[4] The *Raleigh Register* of September 1, 1831, includes this passage about the aftermath of the Turner uprising: "In closing this hasty account, we regret exceeding to state, that Mr. Shepard Lee, an esteemed member of the Halifax Blues was accidentally shot during an alarm, by a brother member of the Corps. The circumstances seem greatly to have excited the sympathy and regret of the circle, where Mr. Lee was best known" ("Insurrection and Murder" 1831).

CRANIAL RECONSTRUCTION 129

or fiery" advocate for murder (Bird 2008, 360), and the "Governor," a "great rogue" who styles himself an enlightened leader and is quasi-literate (Bird 2008, 346).

These caricatures of the two sides of Turner then go on to dissect their various motivations for the rebellion, resulting in a portrait of misdirected mental propensities. These propensities echo racist theories of the brains of Black people that describe an underdeveloped front of the head (i.e., the seat of phrenology's Intellectual Faculties) coupled with an overly developed top and rear of the head, where the Affective Faculties lie. Combe, in 1825, describes a typical "Negro skull" from Africa thusly:

They exhibit much Hope, Veneration, and Wonder, with comparatively little reflecting power. Their defective Causality incapacitates them for tracing the relation of cause and effect, and their great Veneration, Hope, and Wonder, render them prone to credulity, and to regard with profound admiration and respect any object which is presented as possessing super-natural power. (1860, 343)

Overly certain of their future (excessive "Hope") and swayed by religious fervor, Bird's caricatures of Black cognition already see themselves as future rulers. Instead of self-liberation, Brid imagines the splintered Nat engaging in a power fantasy of rape and rank:

The parson . . . got up to make a speech to inflame our courage. There was in his oration a good deal of preaching, with a considerable sprinkling of scraps from the Bible, such as he had picked up in the course of his clerical career. What he chiefly harped on was that greatness of the negro nation spoken of before, and he discoursed so energetically of the great kings and generals, "the great Faroes and Cannibals," as he called them, who had dis-tinguished the race in olden time, that all became ambitious to figure with similar dignity in story.

. . .

"What you say dah, Gub'nor?" cried Zip the fiddler, with equal spirit: "You be king, I be president."

"I be emp'ror, like dat ah nigga in High-ty!" said another.

"I be constable!" cried a fourth.

"You be cuss'! you no go for de best man!" cried Governor, in a heat: "I be constable myself, and I lick any nigga I like! Who say me no, hah? I smash

130 WRITING THE BRAIN

him brain out—dem nigga!" Governor was a tyrant already, and all began to be more or less afraid of him. "I'll be de great man, and I shall hab my choice ob de women: what you say dat? I sall hab Missa Isabella faw my wife! Who say me no dah?"

"Berry well!" cried Scipio: "I hab Missa Edie"—that is, Miss Edith, the next in age, who was, however, not yet thirteen, and therefore but a poor little child. (Bird 2008, 360–61)

In the hands of Bird, Turner's insurrection does not attest to the general's outstanding mental capacities. Instead, it espouses a view of enslaved people as intellectually incapable of self-governance and republicanism. Having read about revolutionary Haiti in an abolitionist pamphlet, these caricatures of slaves—feeling oppressed *only* because a white voice told them to—aspire to become tyrants to gratify a lust for sex and violence. Still, Bird's decision to disassemble Turner's cranium into two halves seems to paradoxically confirm Turner's assertions of superior cognition. These moments in *Sheppard Lee* are as much an attack on Black intelligence in general as an echo of Turner's "uncommon intelligence," which needs to be broken in two to be ridiculed.

Whether a scientifically well-versed citizen believed in the theory of a single human origin for each "race" or denied any relation, the sudden self-liberation of a Nat Turner required denigration. Through the various lenses of racial cognitive hierarchy (Gould 2008, 62–104), a Nat Turner should and could not exist: While his skin is of a "rather bright complexion," he is universally described as "not a mulatto," as a reward notice for his arrest puts it ("Five Hundred Dollar Reward," 1831). There is no "white blood" to justify Turner's brilliance, nor a long history of benign "uplifting": Turner's brain turned genius without genetic or institutional midwifery by its oppressors.

Like Melville's later Black revolutionary Babo, we discover in Gray's description of Turner "a black of small stature, [with] rude face," of all-Black ancestry, "whose brain, not body, had schemed and led the revolt" (Melville 1986, 258). Or in the words of Gray: Turner is "below the ordinary stature, though strong and active, having the true negro face, every feature of which is strongly marked" and discloses a "fiend-like face when excited by enthusiasm" (Turner and Gray 1831, 8). While racial mixing was often used, including in abolitionist works such as Beecher Stowe's *Uncle Tom's Cabin*, to explain Black mental prowess, no such racial crutch could be found here.

CRANIAL RECONSTRUCTION 131

The revolutionaries Babo and Nat find themselves in good company. Certainly, most white readers at the time would have quickly recognized some concerning echoes of Toussaint L'Ouverture, the infamous general who was instrumental in creating the first, and initially prosperous, Black-led republic in the world. L'Ouverture (1743–1803), a devout Catholic inspired by French republican ideals, had quickly become a household name among educated whites in the United States, serving both as a vivid illustration of the horrors of a successful slave revolution and as a lightning rod for radical abolitionists (such as J. R. Beard, who published one of the many popular biographies of L'Ouverture). Like Babo in Melville and Turner via Gray, he was small of stature, of all-Black ancestry—and, by most accounts, brilliant. After he died in captivity, L'Ouverture shared another characteristic with Turner: "Medical men were called in. The head was opened; the brain was scrutinized" (Beard 1853, 277).

Great Brains

In a familiar move, L'Ouverture and his mind were quickly assessed to be "uncommon." The abolitionist Lydia Maria Child famously called L'Ouverture "one of the best and bravest of the negro race" and in "every way so much superior to the other negroes, by reason of his general intelligence and education, his prudence, activity and address, not less than his bravery" (1833, 177). The question of Black genius, embodied by L'Ouverture, was not merely one of *individual* importance but connected to larger questions of the "negro race" and the development of "American" and "African" culture in popular discourse. After all, was it not singular, superior beings that, to a nineteenth-century mind, heralded the general progress of a people? If, as Carlyle had famously proclaimed, "Universal History . . . is at bottom the History of . . . Great Men" (1840, 3), and every major system of thought "is the lengthened shadow of one man" (Emerson 2012b, 170), the absence or presence of Great Men of African descent makes or breaks fantasies of white supremacy.[5]

[5] The term "white supremacy" dates back to the nineteenth century (*OED* dates it to 1824), although it was generally invoked significantly less than in later centuries. The term sees an increase in frequency of over 7,000 percent from 1865 to 1965, for instance, in Google's book database. While the term had little traction in the antebellum United States, it is still used here to adequately reflect the basic outline of then-current concepts of race, as well as indicate the real-life effects (slavery, dispossession, rapes, and killings) of a perceived inferior "race" by a self-proclaimed superior one.

132 WRITING THE BRAIN

Consequently, "phrenological casts" were quickly made of "Toussaint's head," which, according to an English journal (edited by phrenologist Chambers) later in the century "represent [him] as having a skull more European in its general shape than that of almost any other negro." From the worldview of Great Man theory, this suggested that "Toussaint l'Ouverture was not a mere exceptional negro, . . . but that he was only the first of a possible series of able negroes, and that his greatness may fairly be taken as a proof of . . . the negro character" ("Toussaint l'Ouverture and the Republic of Hayti" 1872, 27). If the General Mind of progress expressed itself in Black geniuses as well, perpetual servitude can scarcely be the universe's plan in store for the "race."

This somewhat implicit problem becomes explicit in the writings of the formerly enslaved author William Wells Brown (1814–1884). Now known mostly for his novel *Clotel; or, The President's Daughter* (1852), Brown composed several books that embrace Great Man theory to counter the common stereotype of Black culture and history as one devoid of greatness and characterized exclusively by servitude—a characterization that even Emerson echoes in his abolitionist writings.[6] Works such as Brown's 1847 autobiography, his *The Black Man, His Antecedents, His Genius, And His Achievements* (1863), and his *The Rising Son or, The Antecedents and Advancement of the Colored Race* (1874) expressly counter Emerson's *Representative Men* of 1850—a book that describes societal advancement as the results of nature embodying genius in larger-than-life leaders following certain archetypes, without acknowledging any "Representative Man" of a darker skin tone than that of southern France.

Both Brown's *The Black Man* and *The Rising Son* are emancipatory counterhistories that claim a rich cultural heritage for Black people in the United States at a time when the alleged absence of such history was weaponized to justify servitude and unequal status as symptoms of cognitive inferiority. "We cannot look," Carlyle had claimed, "upon a great man,

---

[6] In his Emancipation Lecture of 1844, Emerson expresses this attitude as follows: "From the earliest monuments, it appears, that one race was victim, and served the other races. In the oldest temples of Egypt, negro captives are painted on the tombs of kings, in such attitudes as to show that they are on the point of being executed; and Herodotus, our oldest historian, relates that the Troglodytes hunted the Ethiopians in four-horse-chariots. From the earliest time, the negro has been an article of luxury to the commercial nations. So has it been, down to the day that has just dawned on the world" (2012b, 297). The same essay speaks of "the more intelligent but precarious hired-service of whites" (compared to Black slaves) while hinting at least at a *potential* equality of cognition: "the intellect,—that is miraculous! Who has it, has the talisman: his skin and bones, though they were of the color of night, are transparent, and the everlasting stars shine through, with attractive beams."

CRANIAL RECONSTRUCTION   **133**

without gaining something by him" (1840, 2). Brown corrects the striking absence of any Carlylean "light-fountains" of Black skin. In addition to short histories of African cultures and (more detailed) treatises on Haiti, as well as brief ethnographies and passages on racial science,[7] both works center around biographical sketches of Black genius. Unlike Emerson for his white geniuses, Brown sings the praises of "Representative Men *and Women*" (as a section is titled in *Rising Son*).[8] The stated goal of these books is to argue that, just as the Anglo-Saxon "has improved his features, enlarged his brain, and brightened in intellect" (Brown 1874, 86), those of African descent are similarly engaged in cranial advancement.

Brown gets explicitly cerebral in these sketches that include Great Men such as Nat Turner ("a martyr to the freedom of his race") and L'Ouverture ("Touissant liberated his countrymen; Washington enslaved a portion of his"). For instance, Phillis Wheatly, via William J. Wilson's estimation in quotation, is characterized as having had a "facial angle [that] contains full ninety degrees; the forehead is finely formed, and the brain large" (Brown 1863, 231), while sculptor Edmonia Lewis's "head is well balanced, exhibiting a large and well-developed brain" (Brown 1874, 465), an aspect she shared with abolitionists Mary Ann Shadd and Jeremiah B. Sanderson, who had a "fine head" (Brown 1874, 539) and a "finely formed, well-developed head" respectively (Brown 1863, 92). Even his *Clotelle* contains a craniological celebration of a slave that seems to anticipate Brown's later works and disclose a long-standing craniological bent in the author's cerebral worldview:

> This slave, whose name was Jerome, was of pure African origin, was perfectly black, very fine-looking, tall, slim, and erect as any one could possibly be. His features were not bad, lips thin, nose prominent, hands and feet small. His brilliant black eyes lighted up his whole countenance. His hair, which was nearly straight, hung in curls upon his lofty brow. George Combe or Fowler would have selected his head for a model. He was brave and daring, strong in person, fiery in spirit, yet kind and true in his affections, earnest in his doctrines. (1864, 57–58)

---

[7] In these moments, Brown even embraces hierarchical racial thinking—likely in an attempt to differentiate between African American and African culture—and echoes claims that African facial features "approximate the beast" because of distance to God and cultural deprivation, not fundamental racial difference.

[8] Emphasis mine.

134 WRITING THE BRAIN

In the mold of Turner and L'Ouverture in the public imagination, this enslaved man has the potential to be a Great Man: "perfectly black" with a "slim" body and "small hands" but an impressive head.

That a vocal Black abolitionist such as Brown would speak fondly not only of Combe, who is now often (merely) described as a fierce proponent of "black inferiority" (Ewen and Ewen 2008, 112), but also of a proto-psychology now rightly associated with scientific racism, may come as a surprise. While some of Brown's goodwill may be based on his personal friendship with Combe (and a belief in Combe's "deep interest in the cause of the American slave"[9]), some of it was connected to phrenology as a medico-materialist framework to more 'objectively' argue for Black liberation. This framework Brown shared not only with Whitman but also with Henry Ward Beecher (John R. Howard, in Beecher 1891, 31) and many of his fellow crusaders against slavery (Hamilton 2008).

"It is easy to see the grounds for scientific scepticism and to apprehend the racist implications of phrenology," Cynthia Hamilton, in her work on abolition and phrenology, cautions. "It is less easy to understand the enthusiasm with which it was embraced by reformers, especially abolitionists" (2008, 179).[10] Indeed, while Combe's early theses on racial inferiority are often critically commented on and used to turn his anti-slavery stance into mere paternalism (Poskett 2014), it seems that Combe's interest in and subsequent travels to the United States, as well as conversations with freedpeople like Brown, changed his mind. Instead of espousing rigid racial difference as he did in his 1825 *System of Phrenology*, Combe's most widely read work, *The Constitution of Man*, began cautiously arguing for potential cranial equality, starting with its seventh edition (of 1836). "Greatly as I respect the character of the Americans," Combe therein states, "it is impossible to approve of their

---

[9] "Although not far from seventy years of age, I found him apparently as active and as energetic as many men of half that number of years. Mr. Combe feels a deep interest in the cause of the American slave. I have since become more intimately acquainted with him, and am proud to reckon him amongst the warmest of my friends. In all of Mr. Combe's philanthropic exertions he is ably seconded by his wife, a lady of rare endowments, of an attractive and engaging manners, and whose greatest delight is in doing good. She took much interest in Ellen Craft, who formed one of the breakfast party; and was often moved to tears on the recital of the thrilling narrative of her escape from slavery" (Brown 1855, 266–67).

[10] Even Stephen Jay Gould in his groundbreaking *Mismeasure of Man* admits to a "warm spot in [his] heart for the phrenologists" and "theory of richly multiple and independent intelligences" that goes against the unitary view of intelligence that *Mismeasure* traces (Gould 2008, 22). Hence, phrenology is not discussed in detail by Gould. Consequently, Gould deemphasizes the shared nexus between craniology and cranial measurement. There is, of course, no clear-cut distinction between cranial measurement for unitary intelligence and cranial measurement for multiple intelligences, as Gould correctly suggests in his discussion of Broca (2008, 129).

treatment of the Negro population" (1836, 40). This disapproval now extended toward theories of innate, immutable racial difference:

> The argument that the Negroes are incapable of civilization and freedom, is prematurely urged, and not relevant although it were based upon fact. The negro head presents great varieties of moral and intellectual development, and I have seen several which appeared fully equal to the discharge of the ordinary duties of civilized men. But the race has never received justice from its European and American masters; and until its treatment shall have become moral, its capabilities cannot be fairly estimated, and the judgement against it is therefore premature. (Combe 1836, 240)

Knowledge of Combe's writings allowed Brown to embrace a monogenic perspective (i.e., the belief in a shared humanity) and argue for a significantly faster-paced achievement of equality than other (pseudo)scientific theories of the day could fathom.

Indeed, at that time, phrenology *is* a Great Man theory, with its various activists and researchers obsessively scouring the busts, skulls, and portraits of famous men (and, *far* less frequently, famous women) to find justifications for the science as well as instructions for societal improvement gleaned from these visages. Brown's writings echo this attitude. In Brown's celebrations of Great Men and Great Women of color, these figures become "representative"—stand-ins and "light-fountains" for a larger population. The appeal of Great Man theory is clear: In the same way that an ignorant French peasant is no indicator of the character of France but Napoleon *is*, an illiterate, abused enslaved person is no "representative" of those of African descent but L'Ouverture *is*, Nat Turner *is*.

Great Man theory and phrenology were wildly progressivist, Lamarckian at times, in their shared understanding of the pace of societal reform. A single mind, they agreed, could shape the destinies of thousands within decades. While both monogenic and polygenic racial sciences of the time (especially in the United States) had an inherently deficit-based understanding of mind (as fallen, degraded), the self-help impetus of the second wave of phrenology encouraged an almost absurd conviction in quasi-immediate societal transformation centered on cranial insight ("know thyself") and the radical leaps and bounds of a whole people propelled ever forward by great brains in their midst. The fact that certain enslaved people are fully equal (or even geniuses)

136 WRITING THE BRAIN

in Combe's scientific gaze allows him to argue for speedy and immediate "justice" for the whole "race"—not (just) a call for benevolence or pity.

A Cranial Case Study

Consequently, when the South seceded from the Union, Brown joined Frederick Douglass and others in enlisting Black men into the Union Army. After the war, he would publish *The Negro in the American Rebellion*, which argues for a throughline from Nat Turner to the heroism of the "Colored troops" in the Civil War, asserting an active history of self-liberation as a driving force of the conflict. In the eyes of many white abolitionists, the Civil War instead proved a prime testing ground for their racialized hopes and worries over cranial improvement in formerly enslaved people. "Colored troops" in particular became a focus of those who both *required* "more brains" from freedpeople and asserted that such was at least a possibility. With their foes across the battle lines, these Combean abolitionists shared an assessment of Black minds as lagging behind in development.

Even rather progressive voices eyed Black troops with some cranial skepticism. T. W. Higginson, mentor to Emily Dickinson, supporter of John Brown, and eulogizer of Nat Turner, for instance, voices his paternalistic ambivalence over the minds of the men serving under him by echoing racial craniology while suggesting mutability in this sketch of a fireside celebration:

> [To-morrow] strangers will remark on the hopeless, impenetrable stupidity in the daylight faces of many of these very men, the solid mask under which Nature has concealed all this wealth of mother-wit. This very comedian is one to whom one might point, as he hoed lazily in a cotton-field, as a being the light of whose brain had utterly gone out; and this scene seems like coming by night upon some conclave of black beetles, and finding them engaged, with green-room and foot-lights, in enacting "Poor Pillicoddy." This is their university; every young Sambo before me, as he turned over the sweet potatoes and peanuts which were roasting in the ashes, listened with reverence to the wiles of the ancient Ulysses, and meditated the same. It is Nature's compensation; oppression simply crushes the upper faculties of the head, and crowds everything into the perceptive organs. Cato, thou reasonest well! When I get into any serious scrape, in an enemy's country,

CRANIAL RECONSTRUCTION   137

may I be lucky enough to have you at my elbow, to pull me out of it! (1900, 18)

With cognition crushed into perception by slavery (a clear echo of the "African skull" theory of depressed "intellectual faculties"), Higginson nonetheless finds an endearing, embryonic "mother-wit" displayed at the "university" of the communal fire.[11] Still, his misremembered reference to Joseph Addison's *Cato* at least frames the targets of his racialized benevolence as potential citizens, republican Catos. To follow such daring reconstructionist suggestions, his fellow Unionists needed more convincing.

Thus, the Union Army quickly began to test Black soldiers' physical and mental fitness en masse—ostensibly to assess their ability to serve, but also with an eye toward the thorny issue of postwar citizenship. One of the largest, and understudied,[12] cases of racial research on the brains of Black people in those years was a study conducted by army surgeons Ira Russell and Sanford B. Hunt. Working on behalf of the US Sanitary Commission and sanctioned by the Army's surgeon general, Russell performed around five hundred autopsies on largely Black soldiers who had fallen in battle or died in camp, around 405 (94 percent Black) of which included removing the brain, weighing it, and noting the racial makeup of the body it came from according to racial gradients from "black" to "1/16th white" to, finally, fully "white." Peter Coffman calls this episode "the greatest example of disregard for human decency and dignity of . . . Black soldiers" by Union medical officials (2015, 47). To add insult to injury, many of these unconsented autopsies were performed at Virginia's L'Ouverture Hospital (Hunt 1867, 174) named after the Haitian revolutionary, who also had his brain removed by white surgeons.

Russell's scholarly reasoning for these postmortem mutilations echoes phrenological justifications, underscoring that, in praxis, Stephen Jay Gould's attempts to divorce it from craniology are futile. Still, it is also a sharp departure from phrenology's celebration of the vast individuality of skulls. Instead

---

[11] Compare this to William Makepeace Thackeray, the famous British novelist and Confederate sympathizer, during his travels in the United States in the late 1850s: "They are not my men and brethren, these strange people with retreating foreheads . . . with capacities for thought, pleasure, endurance quite different to mine. . . . Sambo is not . . . my brother" (1946, 187).

[12] An early exception appears to be Haller's 1971 study *Outcasts from Evolution*, which briefly addresses Russell and Sanford (Haller 1996, 31–34). Schwalm's 2018 study on the supremacist outlook of the Sanitary Commission is perhaps the most extensive one, and also includes a detailed and valuable assessment of the Russell/Hunt study, but only briefly discusses the specifics of the anatomic brain study.

138   WRITING THE BRAIN

of looking for "light-fountains," Russell is on the hunt for a racial *average*, already presupposed to be below the white average. His task is to define a degree of inferiority:

> All the intelligent surgeons agree with me that a thorough knowledge of the habits and idiosyncrasies of the negro are of the utmost importance in order to understand and successfully treat his diseases. . . . Two hundred years of servitude, the implicit obedience required, the exemption from all care and anxiety to provide for the future, the extinguishment of all hope of improvement in his civil or social relations, has produced marked physical and moral effects. Self-reliance and exercise of the will has never been cultivated or formed. . . . His highest ideal of enjoyment has consisted in freedom from toil and the gratification of the lower animal instincts. . . . He is superstitious, and believes in charms and diabolical agencies, and often imagines that he is the victim of some supernatural influence. . . . When under the influence of this hallucination, he becomes indifferent, despondent, and gives up in utter despair, dying without apparent cause. . . . The intelligent physician soon learns that he must treat a negro as he would a child. (In Hunt 1867, 170–71)

With the phrenological organs of Will and Self-Reliance undeveloped and Veneration[13] developed in excess, we are back to Combe's 1825 reading of the "Negro Skull"—and Higginson's "[crushed] upper faculties of the head, [which] crowds everything into the perceptive organs." Like Higginson, Russell notes that the "negro encampment is always a cheerful and chatty place" which he believes is due to the fact that "the negro [is] by nature and education . . . social and gregarious" (in Hunt 1867, 164).

We only get selections from Russell's theories and findings through the final report that Sanford B. Hunt filed with the Sanitary Commission. But in the basics these two men seem to agree: "The intellectual acuteness has been very much blunted by centuries of ignorance and servitude," which makes it "now impossible to define his relative position . . . in the scale of races" (Hunt 1867, 177). Their painful theories bear recording here, as they

---

[13] The Civil War itself, and the role of Black volunteers therein, Hunt consequently distorts thusly: "The existence of slavery was undoubtedly the ultimate cause of the war of the Rebellion. Yet, though it involved the deepest problems of race, it was not in itself a war of races. . . . During the struggle the negro remained passive. His ideas of the struggle were not revolutionary, but religious. He believed and waited, his simple mind filled with grand metaphors. . . . When it was decided by the Government to employ him as a solider, he cheerfully enlisted" (1867, 161–62).

CRANIAL RECONSTRUCTION   139

provide a glimpse into the racial thinking not only of these two men, who likely considered themselves abolitionists, but also of the Union Army, which would soon be tasked with enforcing Reconstruction and Black liberation.

The enlistment of Black soldiers in the war, then becomes "a grand experiment" in Hunt's framing of Russell's autopsies, allowing for an anthropological assessment of Black fitness "on a scale of such magnitude as to render its results decisive" (1867, 164). Working with a craniological framework that also incorporates cerebral density, Hunt claims "that the size and weight of the brain is the measure of its intellectuality,"[14] while also allowing for the quasi-phrenological twist that a "*distribution* of volume" in the brain is likely more influential on an individual level (Hunt 1867, 179). The wealth of data collected by Russell thus allows Hunt to assemble this comparative, ethnographical table (see Figure 13)—which he claims allows for a comparison of Black and white "intellectuality."

Considering Russell and Hunt's explicitly white-supremacist framing of their study—and, of course, the overall folly of brain mass measurements—it comes as little surprise that the surgeons discover confirmation of their biases. On average, as Hunt summarizes Russell's work, a Black brain weighs five ounces less than a White one, mixed blood only begins to have a positive effect at 50 percent "white," and Black soldiers have the highest number of extremely small brains in the study (1867, 182–83).

In their interpretations, however, these men differ from a figure such as the by now notorious racial hierarchist Samuel Morton[15] (whose cranial measurements form the basis for Russell and Hunt's ethnological comparisons). Noting a difference between European and American whites, they add a Jeffersonian twist to the Combean argument:

> Morton's measurements seem to show that during two centuries of servitude the negro brain, if it has not diminished in size, has not increased under the influences of slavery. Therefore the crucial experiment of freedom and education has only just begun. We cannot judge the ultimate capacity of the negro from that which he has thus far manifested. . . . [If] the American is heavier and larger than the European brain [and if] it has enlarged

[14] "Intellectuality," of course, also being a phrenological term.
[15] On Morton, see Gould 2008, 88–100. In 1851, the *Charleston Medical Journal* famously claimed the Pennsylvanian of Quaker stock as a proto-Confederate: "We of the South should consider him as our benefactor for aiding most materially in giving to the negro his true position as an inferior race" (in Gould 2008, 101).

Number of Autopsies.	Grade of Color.	Average Weight of Brain.	Maximum Weight of Brain.	Minimum Weight of Brain.	Brains, 60 ounces and over.	Brains, 55 and under 60 ounces.	Brains, 50 and under 55 ounces.	Brains, 45 and under 50 ounces.	Brains, 40 and under 45 ounces.	Brains, 35 and under 40 ounces.	Brains less than 35 ounces.
		oz.	oz.	oz.							
24	White,	52·06	64	44¾	1	4	11	7	1	.	.
25	¾ "	49·05	61	40	1	.	10	12	2	.	.
47	½ "	47·07	57	37¼	.	2	13	19	12	1	.
51	¼ "	46·54	59	38½	.	2	10	22	11	6	.
95	⅛ "	46·16	57	34¼	.	1	15	50	21	7	1
22	1-16 "	45·18	50½	40	.	.	3	10	9	.	.
141	Black,	46·96	56	35¼	.	5	42	51	38	3	.
405	.	.	.	.	2	14	104	171	94	17	1
Autopsies of Clendenning, Sims, Reid, and Tiedemann, 278	Whites, collated from various sources, .	49½	65	34	7	28	99	97	39	7	1

Figure 13 Ethnographic table of brain weight data collected by Ira Russell (Hunt 1867, 182).

> under our institutions, why may not the negro brain . . . also increase its size? . . . [There] are evidences that the American, in founding a new nationality, has also established a new type of manhood. (Hunt 1867, 184)

This leads Hunt to muse on what might be considered an early, racial take on the "nature vs. nurture" debate:

> As between the two races, the problem is: Does the large brain by its own impulse create education, civilization and refinement, or do education, civilization and refinement create the large brain? This problem might be solved by a series of researches in the weight of brain of poor whites of the south, known as "sand hillers," "low-down people," or "crackers." With them civilization has retrograded. They came of a good stock originally, but have

CRANIAL RECONSTRUCTION   141

degenerated into an idle, ignorant and physically and mentally degraded people. They general aspect would indicate small brains. If they are small it is due to the absence of educational influences. (1867, 185)

Hunt's argument is a fascinating early example of culturally significant dismissal of poor southerners by northerners as racial less-thans as well as a conceptual reversal of the contentious relationship between poor whites and (formerly) enslaved persons in the South.[16] It suggests a slippage into "blackness" as a threat to whites, while simultaneously casting the cognitive abilities of the Black men fighting for the Union as degraded—slavery is to blame, certainly, but the status quo is one of inferiority in any case.

Ultimately, Hunt describes "the negro solider" as fit to serve by highlighting the very qualities slavery apologists used to defend the "peculiar institution": jovial "temperament," loyalty, and an (unthinking) service to his superior. "His only disqualifications," he notes, are an affinity for certain diseases and a "lack of education—perhaps of native intellect—that forbids his attainment of the rank of a commissioned officer" (1867, 186). Read with this conclusion in mind, the thesis of the paper seems to argue for a transition from an agricultural, southern system of bondage into a northern system of second-class citizenship and coercion.

Leslie A. Schwalm rightly considers Hunt's paper part of a "massive effort to ensure that racial ideologies would endure slavery's destruction" by the benevolent reformers of the Sanitary Commission (2018, 658)—"incompetent and disagreeable . . . *hirelings*" if not "a set of foxes & wolves," as Walt Whitman once called them (1863). The reason this report ever saw the light of day was US Army surgeon general and neurologist William A. Hammond, who chose to first publish it in his own medical journal in 1867 and then saw to it being republished two years later in the British journal of the notorious racist James Hunt.[17]

Hunt, the speech therapist of Alfred Tennyson's son, had recently been pushed out of the Ethnological Society of London over his pro-Confederate stance and his virulently anti-emancipatory research. Not only was his last paper presented at the Ethnological Society's conference booed, but

---

[16] See, for example, Brewer 1930, Moss 2003, and Merritt 2017. An illustrative example of how dismissive attitudes toward "poor whites" still coexist with anti-Black racism is the work of Charles Murray, namely, his anti-Black *The Bell Curve* (1994) and his attack on lower (white) classes in *Coming Apart* (2012).

[17] No relation to Sanford B. Hunt. For the reprint, see Hunt 1869.

142　WRITING THE BRAIN

a Black abolitionist "dissected" his faulty racial reasoning so eloquently that newspapers gleefully described the event as an embarrassment for the British supremacist.[18] James Hunt's newly founded journal was created with this experience in mind and dedicated to white supremacy. In seeking to republish the US Army's ethno-craniological study in this journal, Hammond created a transatlantic collaboration between the Union Army's medical leadership and a Confederate-minded British racial purist that uses data from dead Black Union soldiers to argue for their cognitive inferiority.

Figures like Sanford B. Hunt, a Republican long before the war ("Republican State Nominations," 1859), then find themselves in a lineage with those complicit in cutting open Turner and L'Ouverture while allegedly working on their behalf. Hunt and Russell, when contrasted with the vast network of craniological racial theorists of the 1800s that spans the Atlantic (from Morton to Agassiz to James Hunt), were certainly minor players in a larger field of scientific falsity.[19] Still, their case is instructive here, as it articulates a racial outlet for the epistemological problem of material cognition. Russell and Hunt performed an act of theoretical racenorming that would later be literalized in the post-Reconstruction era: a biomedical refocusing that bypasses unresolved questions about the nature of mind and brain by instead policing a racialized, anatomically measurable "intellectuality." This model of cognition comes in gradients and threatens to be a slippery slope into mental "degeneracy." Consequently, Russell and Hunt resurface at the height of the miscegenation panic: Around the time of the founding of the second KKK, their names are suddenly back in print—on the pages of William Benjamin Smith's influential white supremacist work *The Color Line* (1905) and in newspapers across the American South.[20]

---

[18] "We are pleased to observe that a member of the despised race was present to defend its dignity before the assembled philosophers. When MR. WILLIAM CRAFT, an escaped slave, rose and with keen wit dissected the paper of DR. HUNT, he offered a valuable practical commentary upon the assumed inferiority of the negro race" ("The British Association and the Negro" 1863).

[19] For an overview of the truly international nature of this endeavor, see Poskett 2020.

[20] See, for instance, "Negro Brains Small" 1906; "Some Biological Differences Noted" 1909; "Nurture or Nature" 1907. Hunt's notes on "crackers" were, not surprisingly, never reprinted alongside these.

CRANIAL RECONSTRUCTION 143

Realism as Psychometry

While the larger field of cranial measurement in the 1800s is beyond the scope and aim of this book[21]—and has been rather conclusively interrogated by Gould and others—it is worth considering here as an example of a conflation of phrenology, craniology, ethnography, and racial "science" as well as for the cognitive coping it implies. In the eyes of a white intellectual class, cranial ethnography makes the mind a problem for others. The white, northern American brain is enshrined as a perfect specimen—a goalpost for the rest of the world, and for the freedpeople (and poor southern whites) in its own country.

With ethno-craniology supplying a degree of "measurability" that antebellum brain scientists promised but failed to deliver, medical intervention was upscaled from individual to race. Instead of Great Men, a Great Racial Average comes into play. This places individuals on a sliding scale of "intellectuality" that upholds supremacy while also threatening each with the prospect of slippage. Prior to race-normed intelligence tests, a cranial Calvinism of sorts comes into being, with a confirmation of membership among the mentally elect reserved for postmortems. Superiority thus had to be performed, worked on, to demonstrate outwardly what surely must lie within.

The Hunt and Russell study exemplifies two historical developments and underscores their interrelation: the birth of the psychological concept of a unitary, measurable "intellect" and the emergence of the post-slavery racial hierarchy of the United States. In literary studies, early psychological theories[22] and racial hierarchism[23] have been convincingly tied to the aesthetic regime of Realism. Here we note their overlap. Indeed, Realism and early psychology in tandem turned away from cerebral uncertainty and celebrated a resurgence of the idea of the "intelligible" and "coherent wholeness

---

[21] In essence, anatomical psychometry is not a theory about brain function but an attempt to essentialize the complexity of cognition into a single variable. As such, it often presupposes hereditary, generalizable psychological concepts such as "intelligence" where more anatomically included theories such as phrenology tried to tie such concepts to specific brain regions and cranial features. Once psychometric tests, such as the intelligence test, were developed, they naturally used psychological, not anatomical, testing—underscoring that the field's earlier association with craniology was more pragmatic than philosophical.

[22] For a broad sampling, see Sen 2020, Kronshage 2018, Ludwig 2002, and Tyson 1994.

[23] Besides Warren, see Boelhower 1987, Rohrbach 2002, Daniels 2013, and perhaps Barrish 2005. For an early example, see Gross, who observes a lack of "real objectivity" around the issue of race by Realist writers (1961, 5), overlooking, perhaps, the very problem with the potential racial bias implied by that goalpost.

144 WRITING THE BRAIN

of personality" (Bersani 1976, 55–56). Or as Randall Knoper puts it, "the realist novel . . . shared . . . an epistemology and a cognitive aim" with the science of the mind (2021, 2).

In the waning years of Reconstruction, the notion of a psychological Real would collaborate with the externalized performance of intellect in high style to give voice to a culture preoccupied with preserving its intellectuality from (racial) slippage. Richard Menke, who has rightly deemed Realism an "information system," has observed that the genre aligns "real knowledge with neurophysiological embodiment" (2008, 152). One such alignment is along the lines of race. The dawn of Realism in the United States during Reconstruction is a "drama of consciousness" on the stage of a raced concept of cognition (James 1907, 17). It is no coincidence, then, that some of the most prominent defenders of racial psychometrics at the time, including Paul Broca and James Hunt, were speech therapists.

Casting Realism as a cognitive norming tool is not a new insight.[24] In his Berlin lectures, Friedrich Kittler famously reads literary Realism through the medium of photography, noting their shared interest in the visual frame: "Against the backdrop of photography, literature . . . no longer simply produces inner pictures for the camera obscura that . . . romantic readers became; rather, it begins to create objective and consistent visual leitmotifs" (2012, 139). Kittler believes that photography and Realism in tandem create a way of seeing that anticipates film, ultimately leading him to discover a Foucauldian logic of the carceral state emerging. Specifically focusing on the psychometrics of Alphonse Bertillon and his early mugshots, Kittler notes that the increased resolution and fidelity of the literary-technological frame fed modern state bureaucracies' drive to standardize, assess, and punish (2012, 141–45). Clearly, Kittler is recasting Lukács's point about the purpose and detail of Realist style (cf. Lukács 1972), except, of course, that Realist slippage into mere description is to Kittler not a capitalist corruption but Realism's logical conclusion in a technical sense.

Kittler is right, of course. Sanford B. Hunt, for example, would later become a prison commissioner for the state of New Jersey ("New Jersey News" 1879). Still, Kittler is characteristically silent on the explicitly racial foundations of psychometric Realism. After all, the literary-technological logic of norming and surveillance that is emerging in the latter half of the nineteenth century not only coincides with the dawn of imperial ambitions in Europe (and the

---

[24] Bersani, for instance, refers to Realism as a form of "training" (cf. 1976, 52).

birth of the German nation as an ethnostate) but also the ultimate failure of Reconstruction and establishment of a segregationist regime in the United States. Race is not secondary to work such as Bertillon's but preceded it by decades and laid its philosophical and practical foundations.

Realism scholar Kenneth W. Warren (1993) has convincingly argued that the two developments of Realism and racial segregation are not merely historical parallels but mutually constitutive moments. To Warren, Realism gave voice to a conflicted liberal ideology, buckling under the weight of representing injustice as "Real" while arguing for a largely invisible, non-represented solution to it. The crux both Warren and Kittler observe is that Realism involves a psychometric way of seeing that is structurally in cahoots with the established power structures it aims to critique by presenting its highly selective frame as objective, natural, and immutable. There is a conflation of *reading* with *seeing*[25] that Kittler rightly ties to visual technology. However, his specific epistemological trigger—photography—is too broad to adequately and accurately account for this confluence of psychology, race, and Realism.

What Kittler sidesteps in his discussion of Realism, of course, is that its primary thrill for readers lies not in its surface-level "objectivity" but in its attempt at summoning an *experience* of the Real that relies on just enough noise in the textual frame to make a scenario *feel* believable. What is being normed and measured here, then, is not the Real as such (visually echoed by mugshots) but the carefully curated, psychological experience of the Real via a set of literary parlor tricks. Realism's appeal is *virtual reality*, a constituting and positioning of the reader's psyche in a place that would later be taken up by Hollywood's cameras. The applicable technological icon is thus not the photograph but the stereograph. It is the latter-1800s attempt at generating an immersive cognitive effect akin to VR: two pictures (often taken with a double-lens camera), viewed with the right goggles that separate each eye's visual fields, that create a visual doubling (Figure 14) that tricks the brain into fantasizing a three-dimensional image.

Virtual reality, summoned through a bombardment of contradictory cues from the Real, creates an effect of authenticity, of feeling like one is "there." This experience is illustrative of the Realist appeal. Oliver Wendell Holmes describes the effect thusly:

[25] As Bersani states in passing (emphasis mine): "The realistic novel gives us an *image* of social fragmentation contained with the order of significant form" (1976, 60)—i.e., a psychological experience that disguises ideology as unfiltered apperception.

Figure 14 Stereograph of Black Union soldier watched by white onlookers, 1864. The rock formation in the background was nicknamed "Jeff Davis" (New York Public Library).

> The first effect of looking at a good [stereograph] is a surprise such as no painting ever produced. The mind feels its way into the very depths of the picture. The scraggy branches of a tree in the foreground run out at us as if they would scratch our eyes out. The elbow of a figure stands forth so as to make us almost uncomfortable. Then there is such a frightful amount of detail, that we have the same sense of infinite complexity which Nature gives us.... [The] stereoscopic figure spares us nothing—all must be there, every stick, straw, scratch, as faithfully as the dome of St. Peter's, or the summit of Mont Blanc, or the ever-moving stillness of Niagara. (1859, 749)

What physician Holmes celebrates here is the creation of a subject position in art disguised as an authentic, self-actualized experience of the Real—or what William James would come to describe as "the zest, the tingle, the excitement of reality."[26] This is anticipating the trend in Realism that Lukács would come to bemoan, where essence and appearance blend almost fully. Holmes's "mind feels its way into the very depth of the picture" without realizing that this very "mind"-position is authored by the creator of the

---

[26] "Wherever a process of life communicates an eagerness to him who lives it... there is the zest, the tingle, the excitement of reality; and there is 'importance' in the only real and positive sense in which importance ever anywhere can be" (James 1900, 9). William James goes on to apply this observation to Robert Louis Stevenson's fiction, clearly indicating that the reference to "reality" is (also) to be taken in a literary context.

stereograph.[27] This is a fantasy of "a psychology of stable centers" in the act of seeing (Bersani 1976, 59). It's Auerbach's folly: the fundamental belief in a universalist moment in mediated experience. That Holmes finds himself so irrefutably part of the plot, his mind so easily drawn into its narrative, is, of course, due to the fact that the scenario specifically invokes a person like him. The thrilling event depicted here is a rugged encounter with capital-N Nature. This encounter presupposes a white, male perspective to fully realize its effect (cf. Finney 2014). Even this "random moment" is not "comparatively independent of the controversial and unstable orders over which men fight and despair" but is shaped by and reproduces these "orders" (Auerbach 1953, 552).

Of course, Frederic Jameson has likewise demonstrated that a focus on "scene" in high Realist narratives always already incorporates an element of breakage. While a text might invoke what Jameson calls the "personal identity in the ordinary sense," this relationship is still based on style, which, in the absence of actual magic, can never perfectly call a reader into being. Jameson thus identifies an "impersonal consciousness of the present," a sort of empty affect, that marks the Realist text and subsequently incorporates a reader's identity as its object (2015, 25). While Jameson's interest lies primarily in the dialectic exchanges between these two modes of mind, he powerfully asserts the psychological ready-at-handness of the Realist text: It is a consciousness apparatus on standby, awaiting the empirical, measurable subject to propel it into narrative.

"The unquestioned nature of the narrative discourse" of the Realist text, according to Colin McCabe's definition, "entails that the only problem that reality poses is to go and look and see what *Things* there *are*." This world-as-scene not only generates a vague, temporal "consciousness of the present" but also frames and creates a *specific* psyche to interrogate it: "The classic realist text [thus] ensures the position of the subject in a relation of dominant specularity" (1974, 12). Such "dominant specularity"—that is, the intended way of viewing/reading—is always already gendered and raced, though it has to negate both. Or as Warren puts it, even progressive Realists are "wedded to a radical cause which even they could not call by its proper name" (1993, 70). We might again echo Leo Bersani here: "Realistic fiction serves nineteenth-century society by providing it with strategies for containing . . . its disorders within significantly structured stories about itself" (1976, 63) in the form of a

---

[27] On stereographs and literature in a Victorian context, see also Potter 2016.

general "psychic cohesion" (1976, 6). Reconstruction, the ultimate "disorder" of the American mind, is contained in the training regime of narrative (and narrated) psychology.

Such acts of repression, of course, have to summon a doppelgänger. In this case, the optical stereograph has its very own psychometric double: Paul Broca's *stereographe*. Broca was not only an influential speech therapist and neurologist (discovering Broca's area and hypothesizing its role in language production)[28] but also an adherent of racial science and ethno-craniology. *His* stereograph, like the photographic technique, is an attempt to account for the human in three dimensions—in this case, by allowing the anatomist operating the stereograph to trace skull profiles of a subject's head from any imaginable angle to measure cranial capabilities (Figure 15).

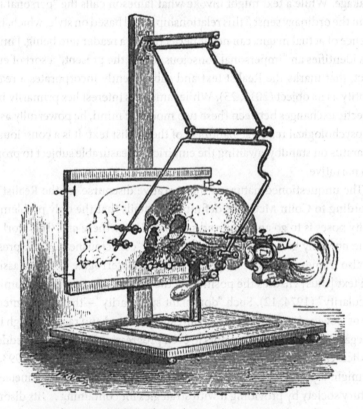

Figure 15 Broca's stereograph (Wikimedia).

[28] For more information on Broca, see Gould 2008, 105–41; Blanckaert 2011.

CRANIAL RECONSTRUCTION   149

In a way, both stereographs are sides of the same coin, interested in subjecthood with the goal of invoking and assessing it. As Bersani puts it, the world of the Realist novel "constantly [proposes] to our intelligence objects and events which contain human desires, which give to them an intelligible form" (1976, 53). Intelligibility and intellectuality—character psyche and reader psyche—are pulling on the same string.[29] Behind stereoscope glasses or mounted onto the *stereographe*, the human head and its contents become readable and predictable: One creates "dominant specularity," and one measures it—psyche and psychometry, generated from subject fictions strapped into an *Aufschreibesystem* that turns the complexities of cognition into legible inscriptions on a piece of paper.

This dual sense of stereography—as creating and measuring degrees of "intellectuality"—is echoed in the reconstructionist literature that was born at the dawn of modern psychology as well as the segregationist psychopolitics of the post–Civil War United States that soon came to rely on race-normed "intelligence tests" to disenfranchise minority populations. The Great Men (and Great Women) of Romance, Gothic, or Sentiment fiction are no longer possible in these texts. There is only one Great Man in Realism proper: the subject position created by the text for the reader (distinct even from a first-person narrator) that cannot be acknowledged as such without breaking the fiction of scenic objectivity. In front of the novel's elaborate, weighty, and well-spoken "consciousness of the present," no character can quite measure up. The Realist novel cannot stutter, though its characters can. Their motivations are measurable, assessable, and ultimately comprehensible in full as "*things* there *are*" (McCabe again). So while a story might argue against racial discrimination, all of its characters constitute an inferior cognition in varying degrees to the presumed white subject position of the reader, whose "mind feels its way into the very depths" of its story and has to be welcomed with open arms and lobes.

High Realism itself, as increasingly invested in modern psychology, not brain anatomy, falls beyond the purview of this book.[30] Still, its relationship with psychometry, subjectivity, and race can already be observed in some early quasi-Realist texts of the day. John William De Forest's 1867 *Miss Ravenel's Conversion from Secession to Loyalty*, famously "realist before

[29] "The exertion toward significant form in realistic fiction serves the cause of significant, coherently structured character" (Bersani 1976, 55). The reader is one such character.

[30] Bersani's study might serve as a stand-in—with the added caution to think "reader" as part of "character" (cf. 1976, 55–59).

150   WRITING THE BRAIN

realism was named" (Howells 1887, 484), for instance, is a prime example of the stereographic gaze. In the Civil War tale, which is ostensibly occupied with reconstructing the mind of the nation by coupling a Union man with a Confederate southern belle, we find several abolitionist statements uttered by Unionist characters about equality, such as: "negro children are just as intelligent as white children" (De Forest 1867, 257). At different moments, however, the reader is tasked with digesting narrative moments like these:

> "Which do you find the most agreeable," she asked, "the white people of New Orleans, or the brown?"
> Colburne was tempted to reply that he did not see much difference, but refrained on account of Miss Ravenel; and, dropping satire, he entered on a calm defence, less of himself than of the mixed race in question. He affirmed their intelligence, education, good breeding, respectability of character, and exceptional patriotism in a community of rebels. (De Forest 1867, 186)

Colburne, a Union soldier and the ultimate "converter" of "Miss Ravenel" toward "Loyalty," here clearly invokes racial mixing to justify "intelligence" in a mixed-race family—measurable at around 50 percent "white," according to his Union colleague Sanford B. Hunt, we might recall.

Still, we get this argument in indirect speech, as opposed to the "just as intelligent" statement presented in direct quotation. What may Colburne have said to make the shared voice of novel and reader summarize it as "good breeding"? What relationship did he draw between nature and nurture, "breeding" and "intelligence"—and why did he explicitly lecture about racial mixing to make his point? The reader, whose mind has "felt" its way into this scene, is unthinkingly tasked with internalizing these concepts. There is no possible outside to the narrative voice, no room for disagreement. In immersive reading, the reader's mind accepts the theory of racial mixing leading to intelligence as being a "calm defense" of formerly enslaved persons as objective reality. It is up to the reader to internalize and justify the conclusions of the impersonal consciousness of the text.

Of course, the general question of "dominant specularity" in the early Realist mode goes beyond this example of bias slipping in and disclosing itself as authoritative truth. And indeed, the novel seems to be aware of that, presenting its readers with numerous brief case studies of doctors assessing the cognition of others, seemingly priming a model for the reader

CRANIAL RECONSTRUCTION   151

participating in its cranial storytelling. There is, for instance, this scene in a field hospital:

> They stopped to examine another man who had been shot through the head from temple to temple, but without unseating life from its throne. His head, especially about the face, was swollen to an amazing magnitude; his eyes were as red as blood, and projected from their sockets, two awful lumps of inflammation. . . .
>   "Fetch him round, I *guess*," whispered the Doctor with a smile of gratification. "Holds out beautiful."
>   "But he will always be blind, and probably idiotic."
>   "No. Not idiotic. Brain as sound as a nut." (De Forest 1867, 300–301)

Or this scene of another doctor holding his grandchild:

> "Very like," said the Doctor. "Very like an oyster. His existence has a simplicity and unity very similar to that of the lower orders of creation. Of course I am not speaking of his possibilities. . . . If you could see the inferior face of his brain, you would be able to perceive even now the magnificent capacities of the as yet untuned instrument." (De Forest 1867, 411–12)

De Forest's doctors presuppose a material mind housed in soft-tissued brains encased in hard skulls. These physiological minds then come in evolutionary gradients, reaching from the "lower orders" to "magnificent capacities." It is up to experts to assess, diagnose, and treat them. This task the reader shares with the doctors.

This leads us to perhaps the most intriguing stereographic reading the novel asks of its audience. It depicts Colburne's rakish competitor Carter, a Virginian soldier and prime example of southern manhood, with a hangover:

> When the Lieutenant-Colonel awoke in the morning he did not feel much like going on a pic-nic. He had a slight ache in the top of his head, a huskiness in the throat, a woolliness on the tongue, a feverishness in the cuticle, and a crawling tremulousness in the muscles, as though the molecules of his flesh were separately alive and intertwining themselves. (De Forest 1867, 44)

152 WRITING THE BRAIN

Depending on the "personal identity in the ordinary sense" (Jameson again) that finds itself thinking here, this mockery of Carter might fall flat. To the craniologically well-versed, however, it discloses a white southerner's slippage into Blackness: His "upper faculties of the head" are "crush[ed]" (Higginson again) like that of a (former) slave, he becomes as "wooll[y]" as a "picaninny" caricature, and even the outer layer of his skin ("cuticle") turns feverish—perhaps indicating the presence of the black "pigment excreted from the blood, and interposed between the cutis and the cuticle" that racial scientists hypothesized at the time (Bray 1871, 51).

De Forest's complexly complicit, stereographic narrative here postulates a shared racial understanding with the reader. This understanding, as Warren notes about Realism, is paradoxical: It is a compact between northern, male, ostensibly anti-slavery minds—who hold racial views they know cannot be directly voiced but that act as affect-in-waiting (cf. Jameson) for the right reader. The novel, therefore, has to efface racial theories unless they can be introduced in a socially acceptable manner: to marvel at the "inferior face of the brain" of a newborn, or to ridicule a southerner for "turning black" when inebriated. Any readerly refusal to participate breaks the objective world of the novel. You may not shake your head in a stereograph.

The particular psychometric debate traced here from the 1830s into the age of Reconstruction is thus one of overt violence and complicit advocacy. It is a cultural discourse around race that ends in a simultaneous rejection of racial violence and widespread absorption of the underlying racial theory of a unitary, largely hereditary "intellectuality" that is unequally distributed among supposed human "races." In reaction to the "problem" of Black cognition (embodied by Turner and L'Ouverture), white American brains become codified as superior and simultaneously threatened by "degradation." That the Realist novel is slowly coming into its own at that very time, these pages have suggested, is not a coincidence. It is a hardening of psychometry into the realm of the Real.

While the focus of this chapter lies with American Realist tales—and their explicit connection to maintaining social order through a celebration of intelligence/intelligibility via style—it would be amiss to not at least gesture to the wealth of scholarship of the master of psychological Realism across the pond, George Eliot. A prime example of "psychological analysis in fiction," as a contemporary expert, the psychologist James Sully, informs his colleagues, Eliot's novels are dedicated to "the unfolding of the inner germs of action, the spreading out before the eye those complicated activities of imagination

CRANIAL RECONSTRUCTION   153

and desire, impulse and counter-impulse" (Sully 1881, 388). Scholarship on this theme in Eliot is, of course, extensive—and any true intervention requires more space than allotted here. Still, these final paragraphs hope to briefly connect these threads to the emerging psychometric elements of early Realism. Thus a brief visit to Eliot's *Daniel Deronda* (1876) is in order.

A work of racial advocacy, and Eliot's only novel set in contemporary England, *Daniel Deronda* has been rightly framed as "most critical of the rhetoric of Englishness" (Lovesay 1998, 505). It is an impassioned defense of a racial "other" through an exploration of psyche: in this case, of its eponymous character and the "long-oppressed [Jewish] race" (Eliot 2003, 170). Deronda's coming-to-terms with his heritage is one of the psychological throughlines of the novel. His mentor in the process frames it explicitly as a task of racial psychology:

> The heritage of Israel is beating in the pulses of millions; it lives in their veins as a power without understanding, like the morning exultation of herds; it is the inborn half of memory, moving as in a dream among writings on the walls, which it sees dimly but cannot divide into speech. (Eliot 2003, 444)

There are both nature and nurture components in this eloquent celebration of generational memory, with an "inborn half of memory" connected to Deronda's struggles with assimilation, which Kennedy has read as an allusion to Wordsworth (1979, 124). One certainly notes echoes of the Wordsworthian mind-caves from the beginning of this book.[31] To Deronda, finding himself means traversing the corridors of a brain that is in part *racial exterior* and which must remain illegible for the moment. Parts of Deronda's intellect are thus inherited, an argument put forward by a Jewish character who endorses this fact as a source of pride.

Daniel Deronda, Richard Menke has presciently observed, "fully exploits the possibilities of imagining fictional psychology along the lines of neurophysiology" (2008, 159). These very "lines" are, of course, psychometry and the racial gradient. Like the abolitionist De Forest in his work, Eliot's defense of Jewish identity comes with a logic of measurability and norming.[32] These moments, like in *Miss Ravenel*, are effaced, brief—but necessary for the novel

---

[31] It is perhaps also one of the moments where Eliot breaks the Realist conflation of seeing with writing and acknowledges an exteriority to the subject-position of the book by removing the implied comprehension of this conflation. The writing, in this image, is *seen* as such but not *read*.

[32] Menke seems to suggest as much when he speaks of the novel's drive to "comprehend subjects that fall outside of ordinary awareness" (2008, 160). The idea of an "ordinary awareness" conceptually presupposes a norming of larger populations; without a clear delineation of an average intellectuality,

154 WRITING THE BRAIN

to retain psychological objectivity/believability in a time when such psychology was explicitly raced. As in its American cousin, they come with their own narrative defense that contrasts them with on-the-face prejudice clearly coded as "wrong" by the novel. Here is one such contextual frame, described as "polite pea-shooting":

> Deronda said he had always felt a little with Caliban, who naturally had his own point of view and could sing a good song; Mrs. Davilow observed that her father had an estate in Barbadoes, but that she herself had never been in the West Indies; Mrs. Torrington was sure she should never sleep in her bed if she lived among blacks; her husband corrected her by saying that the blacks would be manageable enough if it were not for the half-breeds; and Deronda remarked that the whites had to thank themselves for the half-breeds. (Eliot 2003, 272)

In this brief moment, the novel's main character—and polite society's secret racial "other"— identifies with and defends the racial abject, the "blacks." Casting himself as a Black man written by a white man, Deronda demonstrates through superior wit his cognitive superiority to the overt white supremacists he is conversing with.

Said wit, however, also obfuscates actual racial attitudes held by Deronda. In a telling moment that relies on Eliot's slippery authorial "we," readers find themselves in a Bermuda Triangle circumscribed by the poles of "impersonal consciousness," invoked author, and the free-indirect prose that echoes *Miss Ravenel's* like attempts at disguising the locus of racial thinking. The passage itself is a stereographic reading of Deronda's psychological reticence and the novel's conflicted racial politics:

> Sprinkle food before a delicate-eared bird: there is nothing he would more willingly take, yet he keeps aloof, because of his sensibility to checks which to you are imperceptible. And one man differs from another, as we all differ from the Bosjesman, in a sensibility to checks, that come from variety of needs, spiritual or other. It seemed to foreshadow that capability of reticence in Deronda that his imagination was much occupied with two

---

no "subject" can "fall outside" it (note, too, the evoked gradients and cutoffs anticipating the standard deviations of the modern IQ test).

women, to neither of whom would he have held it possible that he should ever make love. (Eliot 2003, 267)

Eliot's "we" does a lot of heavy lifting here, positioning the reader and the racial "other" that is the Jewish Deronda in a community of intellectual advancement over less advanced racial "others" at the bottom of the slippery psychometric slope: the San people, here invoked as racial stereotype.

Sander Gilman focuses on this moment to argue that Eliot is a prime example of an upper-class attitude that fuses "a polygenetic view of race [with] a liberal ideology" (1986, 258). Accusing Eliot of arguing against shared ancestry here is overly hasty. Indeed, the specific statement could have come from someone like William Wells Brown, a fervent opponent of the "polygenetic view" but a proponent of the idea of the racial gradient (which, to Brown, can and is being overcome). The statement is, then, a specific allusion to the idea of cognition as a measurable metric that differs across "races" (and time)[33] and informs the "impersonal consciousness" of *Daniel Deronda* as an unquestioned aspect of narrating the psychological Real of Victorian England.

Of course, the notion that Realist fiction is a burgeoning stereographic tool—a tool that has readers assess the minds of characters and that assesses readers' minds in turn—is in one way widely accepted. Many who read these pages will have experienced the consequence of this psychometric logic firsthand. With "reading comprehension" as a specific cognitive skill now tied to the unitary concept of "IQ," today's universities in the United States still select students through dubious psychometric batteries such as the Scholastic Aptitude Test, which regularly assesses students' cognitive potential by having them read and understand works of George Eliot and like adherents to high style.[34] Created at the height of voter suppression through literacy tests, the SAT has a long history of racial bias (see Vars and Bowen 1998). This bias is not coincidental but core to the story of psychometry. The idea of using literature to racially select from the cognitive gradient, this chapter has argued, does not always constitute a misuse of literary equipment in the way many teachers of nineteenth-century fiction might like it to be. Instead, we

---

[33] The San people, as one of the oldest ethnic groups on earth, are clearly evoked here as "modern primitives," suggesting an evolutionary logic over time.

[34] See, for instance, Practice Test no. 7 (College Board 2016, 1), which connects high-speed comprehension of Eliot's elaborate style in *Silas Marner* to a student's potential for success in higher education. Results on the SAT correlate strongly with IQ.

156 WRITING THE BRAIN

see the literature of the time actively *leaning in* to the paradigm of cognitive norming.

The victory of the psychometric gaze is thus also the defeat of what this book describes as "proto-neuroscience." What is left of the biological brain at this point is sheer size—a heuristic waiting to be replaced by IQ. At the close of the century, psychology stepped in where materialist anatomy failed: to read, repair, and optimize the human mind. As an inward stereograph of sorts, psychology presents each subject with a "drama of consciousness," with the "zest" of the Real, that is experienced as internal narrative and would ultimately give birth to talk therapy. Its flipside is embodied by Broca's stereograph: an anxiety-made-technology dedicated to a racial psychometrics of upholding the perceived superiority of the white psychological subject.

5

Rattle-Brained

Insanity as Material Metacognition

My Brain - begun to Laugh -
I mumbled - Like a fool -
And tho' 'tis Years ago - that Day -
My Brain keeps giggling - still

—Emily Dickinson, c. 1862

You, Me, Brain

The Christmas annual of the English illustrated magazine *Belgravia* for the year 1869 featured the promise of a "Rattle-Brained Story" (Ross 1869, 106). It certainly delivered. The piece, also reprinted in the United States a month later (Ross 1870), is a short, quirky body-swapping tale that sketches how an unnamed clerk at a paperweight factory (who is the narrator of the story) switched craniums with a soldier by the name of Copp. After the two had "lost [their] heads" both figuratively and literally (Ross 1869, 108)—being accidentally beheaded by artillery during a quarrel—a man of science manages to reattach their skulls: "The only error he committed was—that he stuck Copp's head upon my body," the outraged first-person narrator informs the readers of *Belgravia* (Ross 1869, 109).

What follows is an odd tale by any measure, with the speaker's body retaining its abilities to reason but finding himself doubled/split between a still-thinking body under Copp's head and his own "head . . . doing all it could to help Copp's carcass to get on in the world" (Ross 1869, 110). Especially Copp's head's "leerings at the opposite sex" bother the speaker's pious body (Ross 1869, 111). They offend the body's sensibilities as much as the body dreads the beatings often received as the result of such digressions. A constant struggle between the baseness of the speaker's body's new head and his

Writing the Brain. Stefan Schöberlein, Oxford University Press. © Oxford University Press 2023.
DOI: 10.1093/oso/9780197693681.003.0006

Figure 16 "Changing Heads," title image (Ross 1869, 106).

body ensues—a relationship inversely mirrored in the other "person" (leering body, pious head). Ultimately, the speaker is mortally wounded in a duel, his dying words musing on the injustices of an undeserving body receiving the credit for its studious head and a more deserving body being subsumed under the identity of a worthless brain: "I cannot die in peace, when I reflect that my body will be buried under the impression that it belonged to Copp, and that the gravestone will bear hated Copp's name." Here, the ultimate, conclusive marker of one's identity—the *head*stone—turns farcical.

What is especially "Rattle-Brained" about this tale is its tripartite narrative setup: It tells the story of an "I" discoursing about and against a "head" (etymology of "Copp")[1] and a body, wondering how they interrelate and to what extent they can be considered part of the "I"—and ultimately realizing how utterly dependent the self is on both. As a "postscript" informs the reader: "Shortly after Captain Copp [i.e., the speaker] was buried, a report became current that there was something wrong with the other gentleman's head" (Ross 1869, 112). While the story could certainly be considered merely escapist reading, it nonetheless illustrates, quite succinctly, a key philosophical quandary of the time. It describes a "self" that has utterly lost its locus.

[1] Cf. Hanks 2003, 365.

RATTLE-BRAINED 159

Figure 17 Comparing legs (Ross 1869, 109).

The self is both embodied and disembodied: "I" lays claim to the physical form (*my* body, *my* head) but finds itself at the mercy of it, in turn.

The piece's illustrations drive home the point, featuring an apt depiction of the conundrum in a pair of legs that seem to belong to the same body but are assigned different ownership. One leg is "mine," and one leg belongs to the head ("Copp's");[2] the disembodied-but-embodied authorial fingers pointing at both underscore their impossible distance from the situation. What comes to the forefront here is the result of half a century of distilling and digesting the lessons of the Romantics through public discourse: If all external nature is mind (in degrees), there really can be no coherent, singular self.

---

[2] In terms of plot, the illustration serves to show the similar calf sizes of the two, a detail mentioned to emphasize how their naked bodies on the physician's table could easily be confused. Given the explicit state of undress of the men, one has to wonder whether Ross has hidden a perhaps slightly inappropriate comparative anatomy joke in this Christmas annual. As the speaker excuses himself: "By the way, Copp had not got large [calves] either" (1869, 109). This conflation of legs with genitals is, of course, an ancient one—with echoes going back to the Hebrew Bible.

160 WRITING THE BRAIN

While the Romantics turned outward and found identity in nature—scientized as every atom potentially constituting a sort of mini-brain—others turned inward and found a lack of it. To achieve this in narrative, the self had to be alienated, turned into a biology that is distinct from the narrator's everyday experience but still comparable in kind. One's physical self, body and brain, had to be "othered" to strip away habitual assumptions and ask anew the question: What am "I"? There is perhaps no better way to illustrate this point than to speculate: If one were to "inhabit" a new brain, would the sense of "I" shrink or expand? Luckily, a physician by the familiar name of Robert Montgomery Bird had already performed such an experiment in much more detail: his 1836 *Sheppard Lee*, the first major body-swapping tale of the century, arguably laid the cultural groundwork for a piece like Ross's.

Metempsychosis as Metacognition

Bird seems like the perfect candidate for such a mind experiment. As a young man, he had extensively read philosophy of mind at Philadelphia's Germantown Academy (a preparatory school)[3] before he entered medical school at the University of Pennsylvania in 1824. Later, he opened a private practice and became a professor at Pennsylvania Medical College. Benjamin Rush had taught at the University of Pennsylvania until his death in 1813 and his influence still loomed large over its anatomists when Bird took up studies there: anatomy, surgery, and materia medica were taught almost exclusively by Rush's former students and colleagues, such as William Edmonds Horner, Philip Syng Physick, and John Redmond Coxe.[4] Many of these figures were also active in the Phrenological Society of Philadelphia, a group primarily

---

[3] The principal of the Germantown Academy, in a recommendation letter for Bird, specifically mentions his reading of Dugald Stewart's *Philosophy of Mind* (Johnson 1824). Stewart's work summarizes the philosophical debate over the nature of mind—from Plato to Descartes to Hartley—but refrains from engaging in a purely materialist argument. Commenting on such concepts of the mind, Stewart argues (after summarizing them): "If we proceed no farther than facts for which we have the evidence of our own consciousness, our conclusions will be no less certain, than those in physics: but if our curiosity leads us to attempt an explanation of the association of ideas, by certain supposed vibrations, or other changes, in the state of the brain; or explain memory, by means of supposed impressions and traces in the sensorium; we evidently blend a collection of important and well-ascertained truths, with principles which rely wholly on conjecture" (Stewart 1792, 11). Bird was also well read in chemistry (Johnson 1824).

[4] In a lecture at the Medical College, Bird specifically lists Physick and Rush as prototypes of the heroic man of medicine, who seeks incremental truths and not the short-lived fame of quacks (Bird 1841, 14).

RATTLE-BRAINED   161

dedicated to Rush's definition (and coinage) of the term. The "phrenology" that Bird would have been exposed to at the University of Philadelphia was primarily a rationalist, brunonian,[5] materialist account of the mind, and not interested in reading head bumps.[6]

New England, and specifically Pennsylvania, was then the hotbed of so-called heroic medicine (Sullivan 1994). Perhaps due to its apparent success as a treatment method during the yellow fever epidemic of 1793, the system of medicine favored by Rush attained a degree of respectability and cultural dominance in the first half of the nineteenth century that it failed to achieve elsewhere in the world (Sullivan 1994). These theories combined a "mechanistic thrust" (McGovern 1985, 40) with a striking degree of optimism when it came to treatment. Under the eyes of heroic medicine, diseases were considered a problem of unequal organ activation, whether such an issue affected the liver or the brain. Rush thus considered "madness to be a disease of the blood-vessels of the brain" (1812, 25).

As with a water pump or a chimney bellows, there was an optimal amount of activity for an organ, and too much or too little of it (i.e., too much or too little blood flow) meant disorder and disease. To produce thought was thus physiologically comparable to, say, lifting an arm or moving a leg. It was a bodily effect produced by blood flow to tissue. The brain, thus, became a thinking muscle and mind an effect of using it. While such deterministic notions also bubbled behind much of the British (proto-)neuroscience of the day, its proponents were certainly less upfront about it, making the literary reception of Rushean brain science a perfect case study of the effects of brain-centric concepts of mind. It is no wonder, then, that when mind-materialist Joseph Priestley fled British persecution, he was not only extended a warm

[5] Specifically, Rush's neurophysiological theses are an adaptation of Scottish physiologist John Brown's (1735–1788) generalized system of medicine. According to Brown, "fibers [in the body] were 'excitable.' Life was hence to be understood as the product of the action of external stimuli upon an organised body—life, pronounced the Brunonians, was a 'forced condition.'" Diseases were then either characterized by too much or too little excitement (Conrad 1995, 395)—processes Rush and his fellow anatomists approached through blood flow.

[6] At the time, the term would not have been associated with the somewhat carnivalesque psychology that Gall's work would give rise to in the mid-nineteenth century. Rush himself was critical of Gall's work and coined the term before the craniologists came to use it. As Rush himself puts it: "From the science to which I have given that simple name [phrenology], I reject the metaphysical ontology of the Schools, which has for its objects the properties of Spirit, or Being, in general, and shall confine myself exclusively to the history of the human mind" (1981, 404–5). In his lecture, Rush goes on to define this materialist science of the mind as (a) the most important science, (b) a purely factual science, (c) an intelligible science, and (d) the most helpful science.

162 WRITING THE BRAIN

welcome in Philadelphia but also promptly offered a professorship at the University of Pennsylvania.

Sheppard Lee seems occasionally to make punning reference to Bird's own time at Rush's university, having its various characters occasionally be "physicked"—a by then archaic phraseology (according to the *OED*) and an apparent nudge at the famous surgeon Syng Physick, Bird's former anatomy teacher and president of the Phrenological Society since 1822 (Bethune-Baker 1901; Foust 1919, 24). The pun would have been obvious to erudite Philadelphians at the time, as the professor's curious name had attracted jokesters for decades. Even the *Medical and Surgical Reporter* excerpted some of it:

> Sing Physic! Sing Physic! for Philip Syng Physick
> Is dubbed Dr. Phil for his wonderful skill;
> Each sick phiz he'll physic, he'll cure every phthisic,
> Their lips fill with Physic, with potion and pill. . . .
> ("Curious Epigram on Philip Syng Physick" 1859)

While this "Dr. Phil" wrote no books and left no account of his lectures,[7] he would have had a particular interest in the brain, even aside from his engagement with the society.[8] Physick had worked with his personal friend Benjamin Rush during the yellow fever outbreak of 1793, and he performed a postmortem to become one of the first doctors to describe yellow fever virus encephalitis (Powell 1993, 16). Like Rush, Physick's medicine was blood-centric, equating flow to a certain region with activity, meaning that yellow fever could be cured through bloodletting, which reduces an organ's (in this case the brain's) overactivity; the same approach was used by Rush in his psychiatric practice, which relied heavily on phlebotomizing (bloodletting) and sensory deprivation.[9] The specifics of what the soon-to-be

---

[7] A writer for the *Medical Bulletin* remembers: "His style might be said to have been monotonous at times; but his observations were so concise and his conclusions so invincible that a great impression was made on his hearers. . . . These lectures were Physick's most conspicuous contribution to surgery, but they, unfortunately, were lost after the telling, as they were never published" (Harte 1906, 343–44).

[8] George Combe, one of the most prominent promoters of Gall's craniology, even dedicated his *Essays on Phrenology* to Rush, likely in an effort to garner legitimacy by association.

[9] In order to increase activity, Rush relied on heat—even when it came to his own thinking. In his autobiographical writings, Rush states he "used to excite [his] mind by encreasing the heat and blaze of [his] fire in winter" (Rush 1948, 92). Rush even credited fevers and fever dreams with several key changes of mind that he had experienced in his life (1948, 94, 99).

physician Bird would have learned from his professor Physick when the time came to cut open the head of a medical specimen remain unknown. Still, he would likely have encountered the lobed tissue contained within primarily as a physical organ that produces mind and can malfunction through over- or underactivity. Given that he was schooled in the English materialists at Germantown Academy, one can certainly imagine how his surgical encounter with a brain under Physick might have set off a train of thought that eventually gave birth to *Sheppard Lee*.

Sheppard Lee, consequently, is a novel about the material paradox of "inhabiting" one's brain. It consists of a series of sketches of members of Philadelphia society whose bodies are taken over by the eponymous deceased Jerseyman and satirized from within. Initially meeting his demise from an accident during a treasure hunt, Lee performs metempsychosis, with his mind upon death jumping to the nearest recently deceased individual and inhabiting that person's body, be it that of a plantation slave, a benevolent Quaker, or his nemesis, a quarrelsome "squire" living next door to him (and the first person he turns into). The idea itself initially does not seem all too far-fetched to the simpleminded[10] Lee:

> Why might I not, that is to say, my spirit,—deprived by an unhappy accident of its natural dwelling,—take possession of a tenement which there remained no spirit to claim, and thus, uniting interests together, as two feeble factions unite together in the political world, become a body possessing life, strength, and usefulness?
>
> As soon as this idea entered my mind (or *me*, for I was all mind), I was seized with the envy that possessed me when I first met the squire shooting over my marshes. (Bird 2008, 52)

Upon switching bodies, Lee believes he is "all mind," not body. He suddenly understands himself as a literal, Humean "animal spirit" now inhabiting the corridors of the squire's brain. Lee has turned himself into a disembodied entity, an authorial voice devoid of locus, torn between two identities but liberated from biology. He is Sheppard Lee *distilled*—his own "fine ethereal principle," as Romantic chemists would have put it (Davy 2011, 74). Or at least so he thinks initially.

---

[10] He describes himself, in a rather Rushean way, as a man with "little energy or activity of mind" (Bird 2008, 10).

164   WRITING THE BRAIN

Lee's peculiar body politic becomes increasingly problematic as the novel progresses since he not only comes to identify and care for his long list of new bodies (though not enough to shelter them from sudden death) but also notices a change of mind—quite literally. He is no longer a *guest* in the squire's body and brain, but begins to feel the other man's feelings and think his thoughts. The distinction seemingly made by Lee between *being mind* and *having a brain* slowly collapses:

> To account for my forgetfulness of this important transformation, I must relate that, although I had acquired along with his body all the peculiarities of feeling, propensity, conversation, and conduct of Squire Higginson, I had not entirely lost those that belonged to Sheppard Lee. In fact, I may be said to have possessed, at that time, two different characters, one of which now governed me, and now the other; though the squire's, it must be confessed, was greatly predominant. Thus, the moment after the transformation, I found myself endowed with a passion for shooting, as if I had had it all my life long, a buoyant tone of mind, and, in addition, as I by-and-by discovered, with somewhat a hot temper; none of which had ever been known to me before. (Bird 2008, 59)

If one were truly to become someone else's brain, why wouldn't one become said person, Lee ponders—and only a darkly humorous chain of deadly mishaps (requiring him to metempsychosize again and again) allows him to ultimately sign his account "Sheppard Lee."

We can certainly hear the philosophy Bird had been reading when Lee thus repeatedly finds himself having to (unsuccessfully) "set [his] new *associations* properly to work" (Bird 2008, 110). Upon body-switching, Lee experiences "a confusion of characters, propensities, and associations," with especially the latter being "imperfect, as if [his] memory had suddenly given way" (Bird 2008, 108). "Memory," "propensities" ("passions" in the parlance of Hartley and Rush), and "associations" are the very tool kits of the brain-centric philosophy/psychology of mind Bird was familiar with. And *Sheppard Lee* seems poised to drive out the last elements of dualism clinging to these notions.

While *Sheppard Lee* is certainly a biting satire of Pennsylvania society (see Schöberlein 2016a), high and low, it is also a satire of the brain-centric philosophy in vogue there. If one truly believes that thought takes place *in* or *through* brain matter—as most major thinkers in the late eighteenth century did—mind cannot but be a material entity. If Hume, for instance, argues that memory is a physical restructuring of the brain and retrieval of memory is spirit channeled through these structures (like wind through a Wordsworthian cave), how could thought ever have the freedom not to be defined by these physical forms? How could one ever claim that the "association of ideas" creates a sort of a transcendent mind, if it is exclusively housed in, and thus defined by, physiological structures?

Sheppard Lee, in visiting brains that are decidedly *not* tabulae rasae, cannot help but think the thoughts, recall the memories, and feel the feelings that these brains afford him. Memory, sympathy, reasoning—all of these elements are supplied by Lee's host brains. What is there, then, that's truly left of Lee's distinct, immaterial self? Next to nothing, it appears. Only a few chapters after claiming "*I* was all mind," Lee now observes, "*My* mind began to misgive me a little," reducing his "I" to the undue abstraction that a tale like "Changing Heads with Captain Copp" would later toy with. One of Lee's new minds even drives him to suicide,[11] underscoring that his doubled identity is at best a lingering scent of a metaphysical Lee that can only ever become a feasible subject position as the voice of the narrator.

These issues, as lighthearted as they seem in *Sheppard Lee*, were far from being mere vaporous "philosophizing" (Bird 2008, 59). They addressed a burning topic of the time: How can one argue, given the increasingly apparent materiality of the mind, that a person is responsible for his or her actions? Indeed, 1830s Philadelphia, Bird's hometown, was one of the hotbeds of the "irresistible impulse" defense, the first modern rendition of the insanity defense in the United States. Two spectacular cases made nationwide headlines after the novel had already come out and had a lasting impact on Edgar Allan Poe (who praised Bird's work),[12]

---

[11] Again and again, Lee, in the brains and bodies of others, attempts to seek help from a variety of doctors, including a "Dr. Doolittle," who diagnoses gout in Lee's current host. This passage almost reads like a parody of Rush, who by all accounts was a bit obsessed with the disease. Even his *Lectures on the Mind* contain half a dozen references to the foot disease. Given Rush's emphasis on the harrowing mental effects of gout, it is only appropriate that Lee, when afflicted by it (as well as by an unfaithful wife), drowns his current body in the Schuylkill River.

[12] For the 1838 cases and Poe, see Cleman 1991. Another good literary example is Walt Whitman's 1845 short story "One Wicked Impulse!," which describes the act of murder by a decent young man in terms of temporary insanity coded in the Gothic language of demonic possession: "Impeded by the fiendish rage which in that black hour possessed him . . . [and with] monstrous imprecations, he twisted a tight knot around the gasping creature's neck, drew a clasp knife from his pocket, and touching the spring, the long sharp blade, too eager for its bloody work, flew open. During the lull of the storm, the last strength of the prostrate man burst forth into one short loud cry of agony. At the same instant, the arm of the murderer thrust the blade, once, twice, thrice, deep in his enemy's bosom!" (Whitman 1846).

166 WRITING THE BRAIN

but the first cases began to be debated in local medical circles by 1831 (Prichard 1831, 53).

Sheppard Lee is, then, also a tale about a "self" compelled into action by a "brain," even though it should know better. The novel's main character *ought* to be a case study of the ideal rational thinker. After all, he is a spirit merely visiting a brain, and hence should be able to distance himself easily from "impulses," petty rages, or jealousy. Still, even when all of these thoughts are not his own, he cannot but follow them: "I could no longer resist the impulse" (Bird 2008, 256), he exclaims at one point, and "I was suffered to obey the impulses" (Bird 2008, 304) at another. Instead of a visitor, Lee becomes an inmate: He begins losing track of time and wanders "in a kind of maze and bewilderment of mind" through Philadelphia (Bird 2008, 304), ultimately ending up in an insane asylum.

The particular instance that lands Lee there is a case of hypochondriasis that follows the exact outline Benjamin Rush provides in his *Medical Inquiries and Observations, upon the Diseases of the Mind.* The mental disease is triggered by a bad case of "dyspepsia,"[13] which makes Lee feel like a "Shetland pony carrying two elephants" (Bird 2008, 391). "Hypochondriasis" (Bird 2008, 392), as defined by Rush, includes not only false beliefs about having contracted diseases but a wide variety of delusions, many of which include becoming animal:

> [The patient] imagines himself to be converted into an animal of another species, such as a goose, a cock, a dog, a cat, a hare, a cow, and the like. In this case he adopts the noises and gestures of the animal into which he supposes himself to be transformed. (Rush 1812, 80)

Bird must have been a particularly attentive reader, unable to resist the comedic potential Rush had in store for him. Besides becoming a hot teapot and an icicle, Lee (in his hypochondriac host) turns pet:

> The next day, however, a still more afflicting change came over me; for having tried to read a book, in which I was interrupted by a great dog

---

[13] Rush considers "hypochondriasis" a "partial derangement," many of which symptoms Lee's current host displays: "dyspepsia; costiveness or diarrhoea, with slimy stools," hyper- or hyposensitivity to temperature, sleeplessness, and delusional beliefs. See Rush 1812, 78–79. *Sheppard Lee* even suggests the specifics here are from "a book," apparently referencing *Medical Inquiries*, with one character stating she "had studied the whole theory and nomenclature of dyspepsy out of a book the doctor lent her" (Bird 2008, 390).

RATTLE-BRAINED **167**

barking in the street, I was seized with a rage of a most unaccountable nature, and falling on my hands and feet, I responded to the animal's cries, and barked in like manner, being quite certain that I was as much of a dog as he. Nay, my servant Epaminondas coming in, I seized him by the leg and would have worried him, had he not run roaring out of the chamber. (Bird 2008, 393)

Unfortunately, Lee's doctor is a homeopath, so according to his "*secundum artem*" (Bird 2008, 392) he initially sets out to fight fire with fire: Under the dog delusion, Lee gets whipped, and when Lee believes himself to be "a chicken, [the doctor] attempted to wring [his] neck, calling [him] a dunghill rooster, fit only for the pot" (Bird 2008, 394).

When all of this fails to help, Lee is straitjacketed and put into an asylum, where he is exposed to the "heroic medicine" Rush favored: "thrust into a sack of pounded ice," "phlebotomized," and provided "a nightcap of Spanish flies" (Bird 2008, 396).[14] This treatment is successful, but not quite in the way it was intended:

This conversion of mine to their own opinion—or, if the reader will so have it, my return to rationality—had a favourable effect on my doctors. They removed (very circumspectly indeed) the strait jacket from my arms; and then, seeing I made no attempt to tear them to pieces, but was, on the contrary, very quiet and submissive, and that, instead of claiming to be Charlemagne the Second of France, I was content to be Mr. Arthur Megrim, of Virginia, they were so well satisfied of the cure they had effected, that they agreed to free me of their company. . . .

In this manner I was cured of hypochondriasis; for although I felt, ever and anon, a strong propensity to confess myself a joint-stool, a Greek demigod, or some such other fanciful creature, I retained so lively a recollection of the penalties I had already paid for indulging in such vagaries, that I put a curb on my imagination, and resolved for the future to be nothing but plain Mr. Megrim, a gentleman with a disordered digestive apparatus. (Bird 2008, 397)

Sanity, here, means performing one "rational" identity over other, "irrational" ones out of fear of being "punished" for nonconforming (one hears

---

[14] Ground-up Spanish fly was used as a blistering agent, a method promoted by Rush (1981, 225).

168 WRITING THE BRAIN

Foucault's gnashing teeth)—the satirical twist being that "plain Mr. Megrim" is, of course, secretly Lee and thus as fictional as the rest.

Whatever brain Sheppard Lee inhabits, he finds it utterly determining this person's existence. There is no struggling against the brain, even if one could step outside of it for a moment; it is the true, material sovereign, and one either submits or goes insane. Whether one believes oneself to be a teapot, Mr. Arthur Megrim, or Sheppard Lee, all identity is thus ultimately fictional, a narrative applied to the experience of being a thinking machine. The name on one's gravestone, as the narrator of "Changing Heads with Captain Copp" would later lament, is a lie that does injustice to the biology thus subsumed under a rationalist moniker.

Insanity as Pop Culture

While *Sheppard Lee*'s take on early asylums is decidedly lighthearted and its double institutionalization—Lee is locked up in the body of an "insane" person locked up in an asylum—may be considered appropriate at least on one level (at least Megrim's brain *is* psychotic), the tale nonetheless raises the question of the socially constructed nature of "mad" identity. With medical assessments of one's internal organ(s) of mind relying exclusively on external observation, what happens in the case of misdiagnosis? What if, in a less metaphysically unfortunate turn of events, a person really was mistaken for somebody like Megrim?

One of the most popular serialized novels of the century had this very twist as its central mystery: Wilkie Collins's *The Woman in White* (1859–1860). This lengthy tale, told by multiple narrators, centers around the devious plot of the devilish Count Fosco to relieve two noble sisters of their considerable wealth. He hopes to achieve this goal by, among other things, having one of them (Laura) institutionalized against her will under the identity of a terminally ill (and strikingly similar-looking) asylum patient (Anne), whose subsequent death would bestow a hefty inheritance on her husband, a lackey of Fosco's.

The normative force of an insanity diagnosis—a diagnosis that turns the sister's very claims to her own identity into proof of her delusional state—is perhaps nowhere as visible as in the subsequent questioning of her doctor. Even upon noticing physical differences between his former patient and his current one, he analyzes the situation in terms of a medical mystery, without even entertaining the possibility of giving credence to Laura's assertions of mistaken identity:

Insane people were often at one time, outwardly as well as inwardly, unlike what they were at another—the change from better to worse, or from worse to better, in the madness having a necessary tendency to produce alterations of appearance externally. He allowed for these, and he allowed also for the modification in the form of Anne Catherick's delusion, which was reflected no doubt in her manner and expression. But he was still perplexed at times by certain differences between his patient before she had escaped and his patient since she had been brought back. Those differences were too minute to be described. He could not say of course that she was absolutely altered in height or shape or complexion, or in the colour of her hair and eyes, or in the general form of her face—the change was something that he felt more than something that he saw. In short, the case had been a puzzle from the first, and one more perplexity was added to it now. (Collins 2013, 334)

A single diagnosis of insanity, the novel suggests, was often enough to indefinitely brand a person as "not believable" and cause the "insane" person to be deprived of all liberties and rights.

The Woman in White wasn't alone in these criticisms. In the decades surrounding its publication, dozens of works like Fanny Fern's *Ruth Hall* (1855), Isaac H. Hunt's *Astounding Disclosures! Three Years in a Mad-House* (1852), E. P. W. and Olsen Packard's *The Prisoners' Hidden Life: Or Insane Asylums Unveiled* (1868), and Lydia Smith's *Behind the Scenes: Or, Life in an Insane Asylum* (1879) supplied often autobiographical tales of women (in rarer cases, men) locked away in asylums against their will, usually by deceitful relatives. These short accounts generally consciously mimicked captivity novels from earlier decades and were inextricably connected to what could be considered the first wave of anti-asylum activism (the author of *The Prisoners' Hidden Life*, for instance, founded an Anti-Insane Asylum Society). At the same time they supplied highly sensational tales full of family secrets, sinister scheming spouses, and mysterious Gothic architecture.

Still, while *The Woman in White* and the autobiographical tales it drew from (and, in turn, helped popularize) certainly had reformist ambitions, it was their entertainment value that rendered them a pop cultural phenomenon. Within months, the figure of the "woman in white" became a sort of mid-nineteenth-century meme, an omnipresent cultural icon. The novel was serialized on both sides of the Atlantic almost simultaneously, in the United Kingdom in Dickens's *All The Year Around*, in the United States in the well-illustrated *Harper's Weekly*. This generated not only soaring sales numbers (with equally successful book runs soon to follow) but an actual *Woman in White* craze that swept both countries. As one of Collins's biographers put it: "The novel was . . . selling in its thousands, manufacturers were producing *The Woman in White* perfume, *The Woman in White* cloaks and bonnets, and the music shops displayed *The Woman in White* waltzes and quadrilles" (Robinson 1951, 149).[15] Theater renditions of the book across the anglophone sphere performed well into the 1870s, and unrelated books came up with similar titles in attempts to cash in on Collins's success (see Figure 18).

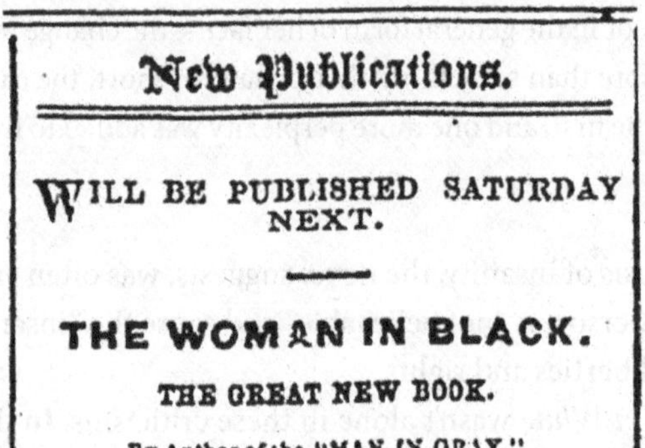

New Publications.

WILL BE PUBLISHED SATURDAY NEXT.

THE WOMAN IN BLACK.

THE GREAT NEW BOOK.

By Author of the "MAN IN GRAY."

A COMPANION TO THE 'WOMAN IN WHITE.'

—BY—

WILKIE COLLINS.

Cloth, $1 50....................Paper, $1 25.

I.

"THE WOMAN IN BLACK." A Great New Novel by the author of "The Man in Gray," and a companion to "The Woman in White," by WILKIE COLLINS, will be published on Saturday next, by T. B. PETERSON & BROTHERS, 306 Chestnut Street, Philadelphia, and will be for sale by all Booksellers and News Agents everywhere. It will be found to be one of the best novels that has been printed in the English language for years and fully equal to Wilkie Collins's "Woman in White." Complete in one large duodecimo volume, bound in cloth for $1 50, or in one volume, paper cover, for $1 25.

Figure 18 Misleading 1864 advertisement for *The Woman in Black* (not authored by Collins) in a Boston paper ("New Publications" 1864).

[15] Even a racehorse by the same name was competing in London within a year of the novel's serial run ("To Be Sold at Auction" 1860). A slight misnomer, considering that the horse was previously advertised as a "*grey* mare, 7 years old" ("The Conjoint St Leger" 1861).

RATTLE-BRAINED **171**

Even public sightings of a woman deemed outstandingly "white" suddenly became newsworthy. In 1868, for instance, one particularly Collinsesque Ophelia was described to an audience more than two hundred miles away for the simple act of boarding a ferry:

> A "WOMAN IN WHITE," says the *Tribune* of Friday:—"Yesterday morning at 11 o'clock a young white lady, having on a white saeque, white dress, white hat, with a white feather and a white v[e]il, white stockings, white gaiters and white kid gloves, and carrying in her hand a white parasol, crossed the Fulton Ferry from Brooklyn, and was 'the observed of all observers.'" ("A 'Woman in White'" 1868)

A young woman, elaborately dressed in all white in the 1860s or 1870s, was clearly seen through the cultural lens of the eponymous asylum inmate(s) of *The Woman in White*. Startlingly, women's fashion trends, as Collins's biographer observed them, seemed to suggest that avid female fans of the novel invited such a reading. It became trendy, it appears, to accessorize in reference to the iconic figure that made the book famous.

This is all the more startling considering that the actual figure of the Woman in White[16] (be it the original "insane" woman or the later kidnapping victim) plays a surprisingly marginal role in the tale: she serves as a tantalizing opening mystery, but she is far from being a fleshed-out character, especially in the latter parts of the story. If the white dress was indeed solely a cultural icon of passivity, victimhood, and infantility, as Sandra M. Gilbert and Susan Gubar have observed,[17] why would so many women choose to wear it as a fashion statement? And why, like the young woman the *Tribune* felt compelled to comment on, would they do it with such an aura of self-assuredness and performativity? Why wear proudly what, in Collins's book, is clearly the asylum uniform of a mental patient?[18]

---

[16] I will capitalize in cases where I am referring to the popular trope, not the novel, or a specific Woman in White in it.

[17] Specifically in reference to Collins's work, Gilbert and Gubar describe the white dress as "snow-image's frosty garment [and] . . . a key term in an elaborate allegory of female vulnerability" and cast the character(s) that wear it in *The Woman in White* as examples of the "Victorian child-woman who clings to infancy because adulthood has never become a viable possibility" (2000, 619).

[18] As Collins puts it, her "dress—bonnet, shawl, and gown all of white—was . . . certainly not composed of very delicate or very expensive materials" (2013, 13). Later in the story, we learn the woman (Anne Catherick) had, indeed, just escaped.

Figure 19 The initial encounter with the Woman in White from the novel's first book editions from 1861 (left: by John Gilbert from the British Sampson Low edition, right: by John McLenan from the Harper's edition).

While Gilbert and Gubar's assessment of the Woman in White trope certainly rings true by the end of the novel (when applied to either Anne or Laura), it neglects the cultural impact of the story's serialization. Certainly, both women end up destitute, somewhat childlike victims, but when Collins's readers first encounter the Woman in White (then Anne), she is far from a helpless maid. In the opening of the novel, the white-clad figure truly is an "extraordinary apparition": a woman walking *alone* at night, in an otherworldly outfit, through the streets outside of London.[19] She is described as cautious but assertive, speaking first and taking charge of the conversation, ignoring probing questions, avoiding niceties, and seeming "quiet and self-controlled" (Collins 2013, 13). She even appropriates a masculine, chivalrous gesture by kissing the male narrator's hand as a token of thanks for helping her find her way.[20] The aura of "strangeness" that surrounds her, here

---

[19] Collins makes clear—as clear, perhaps, as a respectable Victorian writer could—that a woman alone at 1 a.m. on the city streets would have been considered a prostitute: "What *sort of a woman* she was, and how she came to be out *alone* in the high-road, an hour after midnight, I altogether failed to guess. The one thing of which I felt certain was, that the *grossest of mankind* could not have misconstrued her motive in speaking, even at that *suspiciously late* hour and in that *suspiciously lonely* place" (2013, 13, italics mine).

[20] "'Thank you—oh! thank you, thank you!' My hand was on the cab door. She caught it in hers, kissed it, and pushed it away. The cab drove off at the same moment" (Collins 2013, 18).

embodied in the iconography of her odd asylum dress, allows this woman to violate social norms. She transgresses gender and class proprieties with an ease paradoxically available to her by casting herself as, if not flat-out "insane," then at least "strange" (as the narrator repeatedly calls her; cf. Collins 2013, 14).

This self-assertive streak is not a fluke. As the tale builds suspense, not only does the Woman in White's mysterious past draw the reader in, but as a character, she initially retains agency and repeatedly attempts to save the woman who would later take her place in the asylum. She writes cryptic letters of warning, switches hiding places frequently to elude recapture, and even haunts a local graveyard. This Woman in White certainly *is* odd—but this oddness enables an agency and sense of personality (even though she is only observed from a distance) that seem counter to her ultimate narrative fate as well as Gilbert and Gubar's reading of it. If she is seen as the Hodor of this Victorian *Game of Thrones*, the resolution of her backstory's mystery does not temper the agency her otherness allows her when she is first introduced to her audience. And it quickly became a pop-cultural shorthand for the book itself as well as, it seems, for self-empowerment through difference. Eager readers are even informed that the Woman in White is a "dangerous woman to be at large"—a tantalizing label, sure to invite emulation, given that said woman is by that point already established as a potential (though ultimately narratively squandered) heroine.

In the immediate wake of *The Woman in White*'s appearance in the pages of *All The Year Around* and *Harper's Weekly*, the choice of wearing a plain, functional, strictly white outfit thus became a loud fashion statement that indicated membership in the fan culture surrounding Collins's fictional world, as well as serving as an act of rebellion, albeit in moderation. In what appears to be a Victorian version of a subcultural trend somewhere between the *Werther* mania of the late eighteenth century and the goth culture of the late twentieth, these monochrome asylum-like dresses technically violated no norms of propriety but were *just* "strange" enough to make the uninitiated wonder about the wearer's state of mind and to let those in the know recognize a fellow unruly spirit. By donning an outfit like that, Collins's female readership could relive the initial moment of encounter in *The Woman in White* and shock unsuspecting pedestrians while still refraining from clothing choices that would be immediately associated with impurity, poverty, or vice. To meet the Woman in White meant to acknowledge and reflect on her cognitive difference—in the book and on the streets.

174 WRITING THE BRAIN

Given that Collins's literary success with the *Woman in White* rivaled and in certain aspects surpassed that of his rolemodel of sorts, Charles Dickens. It seems only logical then that Collins would follow in his footsteps and make a journey across the Atlantic. In late 1873, the author then generally referred to as "Wilkie Collins, Author of 'The Woman in White'" came ashore in New York and went on an extended reading tour of New England, drawing large crowds while, as observers noted, failing to command his audience with quite the same intensity as Dickens did. Still, the tour would have been considered a success, allowing Collins not only to secure his standing as a first-rate writer of sensation novels but also to mingle with the literary elite of the New World. From Mark Twain to Thomas Wentworth Higginson, from Longfellow to Whittier, Collins was rarely lacking accomplished company (Hanes 2015, 111–17).

Perhaps encouraged by his positive reception in Boston and New York, Collins decided to venture further inland. Before reading to an audience of five hundred in Worcester ("City News" 1874), he was announced to appear on February 5, 1874, in nearby Springfield, Massachusetts (see Figure 20), and the local paper reported "rapid sales" at the local opera house that was to host the event ("Springfield and Vicinity" 1874). The local paper on the next day ran a largely positive review, particularly praising his choice of text, a "weirdly mysterious" story about a nightmare coming true ("Wilkie Collins's Reading" 1874): in this case, a nightmare of an insane woman stabbing the man who had dreamed of her. Echoes of *The Woman in White* are quite pronounced (the tale keeps referencing the woman's "white arms"), but this time she is truly "dangerous to be at large." The audience was thrilled; nobody, the paper states, would soon forget the evening.

Given that a now-famous local resident who lived just a few miles to the north would soon begin to leave an impression on the local townsfolk for, allegedly, refusing to wear anything but a plain, functional white dress (see Figure 21), one might wonder: Were any Dickinsons in attendance? Emily Dickinson had by then been in frequent contact with her quasi mentor Higginson (who met Collins earlier on his tour), was a reader of the *Springfield Republican*, and had attended events at the Haynes Opera in the recent past (Boziwick 2014, 157). *The Women in White* had been on Dickinson's voraciously voluminous list of "secular reading" (Leiter 2007, 47) and one can certainly see why she would have been drawn to Collins's work.

In 1874, Dickinson began her own transformation into a Woman in White, often presumed to be a reaction to her father's death late that year (Philips

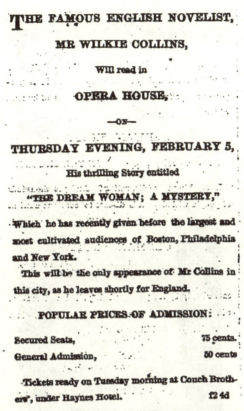

Figure 20 Advertisement for Collins's 1874 reading from the *Springfield Republican* ("Amusements - Meetings - Lectures" 1874).

2014, 794). One of the first extant reactions to this shift in her fashion choices is by Dickinson's posthumous editor Mabel Loomis Todd, who in a letter describes Dickinson as one who "dresses wholly in white, & her mind is said to be perfectly wonderful" (cited in Philips 2014, 794). While the image of the angelic recluse that Todd later created around Dickinson in an effort to promote her works makes it hard to separate fact and fiction, it is startling that Todd's initial description of Dickinson tied her "wonderful" mental state not to her alleged refusal to leave the house but primarily to her white dress.

It is tempting to read this late-life dress choice as, indeed, a fashion statement. For a poet who liked to compose her own cryptic letters and notes, who allowed herself norm transgressions similar to those of Collins's mysterious woman, and who certainly cultivated an aura of "strangeness" around her writings and life, the white dress may have been an elaborate, intertextual

Figure 21 Dickinson's notorious white dress (Emily Dickinson Museum).

piece of performance art, aimed at providing the poet with a sense of "danger" that allowed her to free herself from the most severe societal pressures, while never causing true moral outrage.

At the very least, the public comments surrounding her dress are a testament to the popular culture of her time. Whether Dickinson aimed to be read as Collins's asylum heroine or not, the people around her certainly did just that. With women across England and the United States embracing the figure of the Woman in White as an icon of female cognitive difference, it is hard to imagine the well-read poet would have been ignorant of the cultural framework of her outfit. To look at a woman dressed in monochrome white clothing "not composed of very delicate or very expensive materials" (Collins 2013, 13) during that time meant to consider her mental "strangeness," to treat her as a mystery, a cognitive puzzle to be unraveled. Whether the "Myth of Amherst" (Philips 2014, 794), as the townspeople allegedly termed her, ever intended for such a treatment or not, she certainly engaged with the notion of "insanity" and mental difference in her writings. While the narrative spun around Dickinson's white dress may be just a marketing strategy by her editors, hoping to exploit prevailing cultural notions about "madwomen" in white, Dickinson as a poet was fascinated with "dangerous" states of mind, malfunctioning brains, and the power of the label "mad." And this fascination can be traced from her early years as a student to some of her final pieces of writing.

Psychosis as Metacognition

Where Collins stays securely outside of his protagonists' brains when speculating on their mental state and Bird's satire stops short of the grimmer notions underlying its mind-materialism, Dickinson feels right at home there. Her inquiries begin where those of her male colleagues end: with the notion of insanity, the ultimate tool to dissect the physiological limitations of mind.

That Emily Dickinson's poetry keeps returning to madness is not surprising, nor should it be misunderstood as self-diagnosis (Schöberlein 2015, 46–48). Indeed, the topic would have been hard for her to avoid: Dickinson's life falls squarely into the age of the insane asylum, making it an apt theme for her poetic inquiries. Massive medical institutions aimed at reforming the mentally ill were springing up all around the country and especially in her Massachusetts of the mid-nineteenth century. Dickinson's father even volunteered his valuable time at one of them.

That Edward Dickinson, a local political leader, wholeheartedly embraced the movement toward the institutionalization of the insane (as opposed to neglect or criminalization) is also strongly suggested by one of the essays he wrote in college and by the fact that he served from 1859 to 1864 as a trustee for the Northampton Lunatic Asylum (Habegger 2001, 176; *Annual Report of the Trustees* 1877, 56).[21] In this position, he would have been involved in

---

[21] In the manuscript, written at Yale in 1821 and more than ten years before the first asylum opened in his home state of Pennsylvania, Edward calls the establishment of an insane asylum a "noble object." After didactically recounting the benefits of such a proposal, he closes: "As utility demands, as benevolence urges, and as humanity pleads, for such a retreat, the public ought to lose no time in establishing an asylum when thousands of our fellow beings might be restored to their reason, and

178 WRITING THE BRAIN

key decisions of the institution that he once called "this most important + valuable public charity" (Dickinson 1865). Besides voting on the physician to lead it and recommending improvements to the hospital's facilities, he would also have been expected to "make visits through every part of the hospital" frequently (Kirkbride 1854, 51).

During this time, the Northampton asylum was under the direction of Dr. Pliny Earle, who went to medical school at the University of Pennsylvania, found the influence of Rush there oppressive, and hoped to move away from "heroic medicine" in his practice. Late in his life, he observed of his early days:

> My medical education was received at the school in which Dr. Benjamin Rush had been a professor; and, along with respect, esteem, and affection for the professors at whose feet I sat, I imbibed reverence for Dr. Rush. But his theories of the pathology and his principles of the therapeutics of insanity, and the inconsistencies into which these led him, did not die with their originator. His "Medical Enquirers and Observations" has had a circulation among American physicians more extensively than that of the works of all authors upon mental disorders. . . . No individual authority could overcome the far-prevailing (but, happily, not, as formerly, the all-pervading) influence of Rush in the United States. Even in England his theories are still alive. (Earle 1898, 144–45)

Instead of excessive bloodletting or other semi-torturous Rushean practices that *Sheppard Lee* satirized, the Northampton asylum under Earle's direction followed what was considered a cutting-edge therapeutic program: systematically controlled, health-conducive environments coupled with strict regimens of work, moral and religious instruction, and dietary restrictions, with bloodletting reserved only for truly severe cases—even when still working from Rushean concepts of "cerebral excitement" (1854, 16).

Earle, who in 1863 had accepted the first "professorship of mental diseases" in the United States at the Berkshire Medical College, forty miles west of Amherst, also conversed with Edward Dickinson via letters (Habegger 2001, 715; Wilson and Fiske 1888, 289). Indeed, Edward's interest in psychiatric issues must have been valued quite highly by Earle, who in an 1864 letter expressed "disappointment" over not having seen the elder Dickinson for

which would be an unforeseeable blessing to ourselves, should we ever be [subject] to that same miserable condition" (Dickinson 1821).

two board meetings (Earle 1864). While Edward resigned as a trustee of this "safe & valuable & desirable retreat" in 1864 (in Habegger 2001, 411), the two men remained in contact, with the psychiatrist sending books and Dickinson commenting positively on the doctor's lectures, strongly suggesting that Earle's outlook on psychiatry was generally shared by the poet's father. A prime example of Edward's admiration both for Earle and for "alienism" (as early psychiatry was often called) in general can be found in a letter from 1870, where Dickinson, after congratulating Earle on the positive outcome of a recent investigation over illegally detained inmates, ties the progress of psychiatry to the spiritual ascension of humanity:

> When men in high position, give their names to mountebanks + moon-struck idlers, asking for the investigation of . . . public officers, whose garments are white as snow, + whose lives are a shining light to all about them—it need not surprise, even a casual observer, that depravity has not yet [had] its perfect work—and that there is yet [a] considerable distance between us + the Millennium. (Dickinson 1870)

When Edward Dickinson looks at alienists, he quite literally sees demigods in white. Instead of just being medical professionals, men like Earle have turned into harbingers of the millennium to the poet's father.

How much of his enthusiasm for alienism Edward manifested at the Dickinsons' proverbial "dinner table," so often summoned when trying to reconstruct Emily through her father's discourses, remains unclear. Given Edward's lifelong fascination and activism for this cause and the fact that the family lived at the very center of innovation in the field in the United States, it is likely that the topic of insanity came up occasionally.

Emily Dickinson would not have encountered her father's ideas naively. Having enjoyed a degree of schooling that was, by any comparison, remarkable for her time and learning almost exclusively from textbooks designed for her male peers, she would have understood from a fairly young age what the brain was, what it did, and what it looked like. We know, for example, that Dickinson studied Calvin Cutter's anatomy textbook[22] in 1848 and expressed an interest in it (Lowenberg 1986, 41).[23] "The brain," Cutter informs his readers in his chapter on neurology, "is regarded, by physiologists and

---

[22] On Cutter and Dickinson, see also Baumgartner 2016.

[23] In an 1848 letter to Abiah Root she mentions: "I am now studying 'Silliman's Chemistry' & Cutler's [sic] Physiology, in both of which I am much interested" (Dickinson 1906, 34).

180 WRITING THE BRAIN

philosophers as the organ of the mind." Its function is to bring the rest of the body "under the control of the will" (1847, 231). While bracketing all aspects of mental philosophy, Cutter's work also subscribed to notions of "mental excitement" (1847, 95), leaving open how such "will" ought to function physiologically.

It seems likely that the student whose poetry would later "explor[e] the nature of the mind and the consciousness" would pay special attention to Cutter's neurology chapter (Anderson 1959, 290). Indeed, when we read in Dickinson's poetry of the brain's gyri (what she calls "Corridors," 1999, 188) and sulci (Dickinson's "Groove," 1999, 563), its hemispheric structure (1999, 867), and references to its sponge-like texture ("The Brain . . . will absorb - / As Sponges . . . do," 1999, 269), we hear echoes of the numerous, often fairly gory illustrations in her anatomy textbook. While Dickinson never had any hands-on experience at the dissection table, these textbooks brought her surprisingly close.

A poem like "I felt a Funeral, in my Brain" (Dickinson 1999, 153) might then well be describing the brain's physiology. Indeed, terminologically, an axial cut through the brain (see Figure 22) revealed to Dickinson a "tough corpse" at the center of the brain: a vaguely coffin-shaped corpus callosum, surrounded by black-clothed "Mourners" (little dots representing arterioles). What she describes in these moments is what we would see if "the top of [her] head were taken off" (Dickinson 1906, 315). Dickinson's poetic brain is a physical object, a material thing that has texture, "weight," and mass (1999, 269), strongly suggestive of the poet having visualized the organ hidden behind her own thick skull. Barbara Baumgartner's claim that "Dickinson characterizes intellectual processes, emotional turmoil, and poetic composition as embodied experiences within the brain" thus holds up (2016, 72).

We also learn that Dickinson's brain was indeed an "Arc of White": She would have read in Cutter that it was covered by a protective layer called "arachnoid, or spider's web membrane" (1847, 227). One of her many spider poems, "A Spider sewed at Night" (Dickinson 1999, 456), could then be read as a piece about thought itself, instead of merely depicting the act of writing. The poem is generally read as presenting the author as a "determined craftsman" who appears, oddly, to be composing in the dark of night—or, as Barton Levi St. Armand tries to make sense of it, "by the glare of his own materials" (1984, 38). Instead, it might depict a *mental* weaving even more crucial to Dickinson's artistic project:

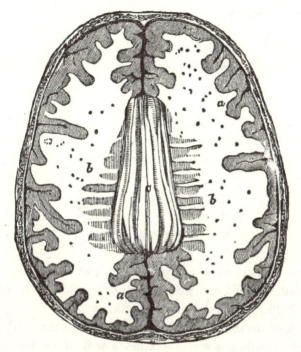

Figure 22 Illustration of an axial section (Cutter 1847, 228).

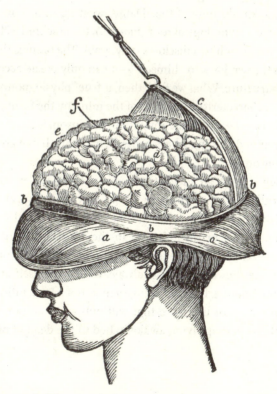

Figure 23 Removal of dura mater (Cutter 1847, 227).

182   WRITING THE BRAIN

A Spider sewed at Night
Without a Light
Opon an Arc of White -

If Ruff it was of Dame
Or Shroud of Gnome
Himself himself inform -

Of Immortality
His strategy
Was physiognomy -

With the knowledge that the brain is covered by a white, web-like layer called the arachnoid, we could very well be witnessing an idea or an image being born in the hidden corridors inside the poet's cranium.

Even in total darkness, the spider—perhaps the brain, perhaps a specific thought crawling across it—excites a web of mental connections in a self-contained system that allows for little outside interference. The brain in Dickinson's telling second stanza "inform[s]" itself; like the form of a "Ruff" that speaks only of the "Dame" wearing it, or how a "Shroud" outlines a "Judge" (etymological root: *gnomon*), the arachnid web of thought (Dickinson's "Arc of White") discloses only brain. The brain is the "himself" that is always thrown back to "himself"—it can only create according to its own physical structure. What we find, then, is true "physiognomy," the "art of discovering . . . characteristic qualities of the mind by the form of the body," as Dickinson's *Webster's* puts it. Thought, to Dickinson, is a veil thrown over the brain—seemingly hiding what's beneath, but really being given shape by it and reflecting on it.

Poetry, then, is not an idealist endeavor—there are no overarching, ephemeral truths to be arrested in verse—but a biological process "of the body." Even in its pursuit of "immortality," the mind cannot leave its embodiedness. It can shroud itself in the "Ruff" of words but never escape its nature. In a Heideggerian hermeneutic loop of sorts, the brain of "A Spider Sewed at Night" always returns to itself and its finite, physiotemporal boundaries. The poem, then, is not a Romantic piece about the immortal nature of literature that will outlive its author. Instead, we find the speaker—perhaps lying awake in bed in the dead of night—being haunted by thoughts of mortality crawling across his or her mind on spidery legs. Knowing the brain for what it is, Dickinson's speaker cannot mentally command "Immortality." That is just not a feasible "strategy" for a physical organ.

The troubling consequence of this for those raised in a religious household such as Dickinson's seems obvious: If will and mind are contained in a "little bone box scarcely eight inches long" (Anderson 1959, 303), then where is the soul? While Cutter's anatomy book could ignore this idea (the word "soul" is only mentioned once), many of Dickinson's poems, such as "This is a Blossom of the Brain - " (1999, 449), appear to address this existential question head-on:

This is a Blossom of the Brain -
A small - italic Seed
Lodged by Design or Happening
The Spirit fructified -

Shy as the Wind of his Chambers
Swift as a Freshet's Tongue
So of the Flower of the Soul
Its process is unknown.

When it is found, a few rejoice
The Wise convey it Home
Carefully cherishing the spot
If other Flower become -

When it is lost, that Day shall be
The Funeral of God,
Opon his Breast, a closing Soul
The Flower of our Lord -

The soul, in this poem, is an outgrowth of the brain. It comes from a seed "lodged" within the nervous matter contained in the skull. Engaging the overt materialist imagery directly, it is difficult to keep reading this poem as about embracing the "Holy Spirit" (Farr 2005, 208), as much commentary would have it, or the "grace of god" (Mitchell 2000, 151). In this poem,

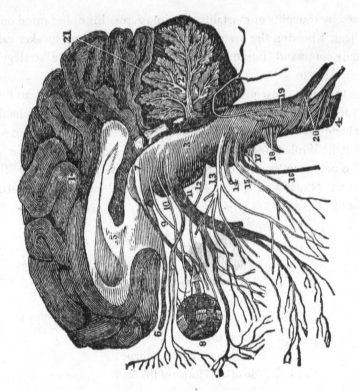

Figure 24 Illustration of a sagittal section (Cutter 1847, 242, turned 90 degrees counterclockwise).

"that from which any thing springs" (*Webster's*, s.v. "seed") is, indeed, inside the brain as a physical object. It is unavoidably a posteriori to the brain and nourished by it to blossom. What theological dangers this materiality of the soul entails are fittingly underscored by Jonathan Edwards's comment on the issue:

> I would know whether this material soul keeps with [the body] in the coffin, and if so, whether it might not be convenient to build a repository for it; in order to which I would know what shape it is of, whether round, triangular, or four-square; or whether it is a number of long fine strings reaching from the head to the foot; and whether it does not live a very discontented life. I am afraid when the coffin gives way, the earth will fall in and crush it. (1840, 58)

Edwards addresses here, in similarly stark imagery, an issue Dickinson also wrestles with: If we begin to equate soul and brain, what happens to the Christian promise of resurrection and a heavenly afterlife?

The brain, in Dickinson's poem, creates and sustains the soul; the soul, in turn, is a natural outgrowth of the brain. Just as a plant without soil wilts, a soul succumbs when the brain fails or ceases to nourish it—an image reminiscent of root-like cranial nerves flowing down below the floral arbor vitae structure of the cerebellum in Dickinson's medical textbook (see Figure 24). At this point, one can only wonder what effects such illustrations must have had on this keenly perceptive poet-student. A "soul," they seem to propose, is an object that follows the logic of nature, springing from a seed (in this case the spinal cord and the brainstem), then growing outward and, ultimately, dying and disintegrating. The illustration suggests an understanding of the brain that does not serve to elevate the mind into heavenly spheres of gods, angels, and souls. Instead, it brings that "which is supposed to be the seat of the soul" (*Webster's*, s.v. "brain") to the level of vegetation: of growth, fruition, and, ultimately, decomposition into soil—a soil that in Dickinson's world both enables the death-bound beauty of summer and, in turn, embraces the corpses of many of her close friends and family members. Dickinson's soul-"Flower" (1999, 449) will certainly be "crush[ed]" when "earth will fall" on it (Edwards again).

With the soul a mere creation of the spongy organ contained by the "repository" that is the skull, the core of Christianity's promise is no more, and Dickinson's speaker in "This is a Blossom" can pick the metaphorical flower to put in the coffin of God, propped up for his funerary rites. Surely, the flower withers (it closes), leading the speaker to recognize God and his promises of immortality as what they are: "a Blossom of the Brain." Whether belief is natural ("Design") or learned ("Happening"), an anatomical look at the lobed structure that produces belief will be its end. We know that from "the spot" that once held the flower nothing new will grow.

With mind, will, and soul all contained in the vegetal "Corridors" of the brain, Dickinson observes her universe shifting away from a metaphysical mind-body dualism—entertained by many of her former teachers, such as Edward Hitchcock—toward a more materialist understanding of mental life. As we have seen in Dickinson's "A Spider sewed at Night," emotion and sensation become physical processes without the necessity of a spiritual component when the soul is reimagined as matter. Poems such as "The Brain, within it's [*sic*] Groove," "I felt a Cleaving in my Mind -," or "I've dropped my Brain - My

186 WRITING THE BRAIN

Soul is numb -" read as so forcefully biological in their descriptions of emotion that they can easily lead someone like Lyndall L. Gordon (2014) to hear Dickinson describing an epileptic attack with medical rigor: Dickinson talks of veins, nerves, and instincts, and imagines the physical thought traversing the brain. Still, while Gordon's analysis seems to ignore the medical knowledge of Dickinson's time and presumes the poet would have just sensed what the neurological underpinnings of, say, a grand mal seizure are, it does point us to the surgical precision with which Dickinson discusses sensations.

It seems all too simplistic, then, to see the poet as merely responding to pathologies or absorbing others' discourses, leaving her reacting to an undisclosed illness, to her father's anxieties, to a fear of going insane—all from a defensive standpoint at best and a passive one at worst. Besides some of the questionable data this would yield, such a biographical heuristic would also ignore Dickinson's ability to encounter medical concepts as what they are: theories. In this regard, Hiroko Uno's analysis that Dickinson was "arguing or experimenting with theories about the relation of science and religion using her learning from textbooks or lectures" (1998, 107) rings true. Instead of the poet grappling with a specific biographical event and grasping for words and metaphors to describe it, we might instead (and more productively) see her getting to the root of what, materially, thought is. Dickinson does so by taking her inquiry to the extremes of the human mind: madness, despair, terror, grief, and death. A poem like "I felt a Funeral, in my Brain" (Dickinson 1999, 153), for example, seems to describe the details of a literal mental breakdown (on an emotional and experiential level) while also keeping in line with the medical knowledge available to her:

> I felt a Funeral, in my Brain,
> And Mourners to and fro
> Kept treading - treading - till it seemed
> That Sense was breaking through -
>
> And when they all were seated,
> A Service, like a Drum -
> Kept beating - beating - till I thought
> My mind was going numb -
>
> And then I heard them lift a Box
> And creak across my Soul

RATTLE-BRAINED 187

> With those same Boots of Lead, again,
> Then Space - began to toll,
>
> As all the Heavens were a Bell,
> And Being, but an Ear,
> And I, and Silence, some strange Race
> Wrecked, solitary, here -
>
> And then a Plank in Reason, broke,
> And I dropped down, and down -
> And hit a World, at every plunge,
> And Finished knowing - then -

Looking at this poem strictly from the medical perspective of her time, Dickinson is describing an incident of "disassociation," or a morbid "association of unrelated perceptions, or ideas" impeding the ability of the "mind to perform the operations of judgment and reason" (Rush 1812, 257). In this case it is a "paroxysm" "excited by the sensible object" (Rush 1812, 257) of a funeral procession and conflated with an internal state of terror. There is movement ("treading") and pulsating sensation ("beating - beating") as the overexcited nerves struggle to compensate. Following an undescribed causal event, images occur of increased mental activity accompanied by what reads like possible visual and auditory hallucinations and ultimately a complete mental disconnect from reality ("Finished knowing - then") with little hope of recovery ("Reason, broke").

While such a short assessment does not come close to unraveling the intricacies of meaning in this poem, it does help us see how Dickinson frames feeling. There is a distinctly scientific undercurrent running through these pieces that is a sign not of the poet's intuiting deficient mental processes within her brain but of her imaginative deployment of medical theories. When Dickinson analyzes a complex emotion, she does so by observing it with "telescopic eyes" (1999, 237). She gets as close to it as she can through the language and concepts of neurology and alienism. And as she does so, she finds these emotions to be physical processes contained in the organ that is the brain. Grief that approaches insanity is, here, not an abstract longing, a sad gaze, or a tear on the cheek—it is a material event in the brain. When it comes to "Emergenc[ies]" of the mind, Dickinson relies not on "Faith" but on

188 WRITING THE BRAIN

the compact, poetic "Microscopes" that she creates out of the science she has read (" 'Faith' is a fine invention," 1999, 95).

Having approached the "nature of the mind and . . . consciousness" with thought experiments like these, we find that another problem arises: If the brain is the mind, to what extent can it be "under the control of the will" (Anderson 1959, 290; Cutter 1847, 231)? And what happens to the "self" when the brain is diseased, its reason "snapt"? "The first Day's Night had come -" (Dickinson 1999, 195) seems to address exactly these issues. Assuring us in its first two stanzas that the "Soul" is subservient to an "I" and made up of "strings" that can be "snapt" and "atoms" that can be "blown," the poem makes us witness what happens when this organic relationship is put to the test. After depicting the resurgence of what appears to be a repressed horror that deeply unsettled the speaker's being, the poem concludes:

> My Brain - begun to laugh -
> I mumbled - like a fool -
> And tho' 'tis Years ago - that Day -
> My Brain keeps giggling - still,
>
> And Something's odd - within -
> That person that I was -
> And this One - do not feel the same -
> Could it be Madness - this?

In a move that Charles R. Anderson calls Dickinson's "self-behind-the-self," the poet treats us to the observation of a mind both from within and from a distance, allowing the speaker to rationally reflect on his or her own irrationality (1959, 306). Taking a stock variation of the "lunatic"—the giggling fool (cf. Rush 1812, 338)—the speaker describes him- or herself as at the complete mercy of a dysfunctional brain. In a telling synecdoche, the lunatic becomes his or her brain ("My Brain keeps giggling," instead of "I keep giggling").[24] While the poem starts from the proposition that an "I" wills and manages a physical, emotive center referred to as the "Soul," it ends with the realization

---

[24] A biographical echo of this can easily be heard in the account Dickinson gives of her brother's coping with the violent death of a friend: "Austin is chilled - by Frazer's murder - He says - his Brain keeps saying over 'Frazer is killed' - 'Frazer is killed,' just as Father told it - to him" (1906, 203). As in the poem, the person afflicted with a traumatic loss becomes his or her brain—and can do little to put it back "under the control of the will" (Cutter 1847, 231).

that said "I" is the mental prisoner of a "Brain." Instead of a will, a self, or a mind rationally managing the speaker's being, this peculiar organ appears to be in dictatorial control. It even seems to mock the speaker's terror upon realizing this reversal of power.

Using the trope of insanity, Dickinson thinks through the limits that a biologically constrained mind imposes on the individual. In madness, Dickinson discovers a striking set of questions: What defines one's being, if the actions of the self are determined by a physical organ never quite "under the control of the will" (Cutter 1847, 231)? How can one even differentiate between "I" and "my Brain" if the physical and experimental reality of madness shows its futility? With the acceptance of strictly neural bases for mind and emotion comes the uneasy possibility that self-determination as we commonly understand it might be illusionary. The self is the brain and vice versa. Only through the poetic doubling of perspective (one involved and experiencing, the other observing and reflecting) can Dickinson articulate on what shaky ground the idea of a distinct, non-material "I" rests.

When Anderson argues that "the poet's traditional role of being 'insane,' and his consequent alienation from society, was one she readily assumed" (1959, 306), we should keep in mind that Dickinson's texts rarely relish the sort of Romanticist yearning and melancholia that often accompany such poet-figures. Dickinson does not surrender herself to a "melancholy fit" that "shall fall / Sudden from heaven like a weeping cloud" (Keats 2009, 250) but meticulously analyzes specific breaks in reason and brain obstructions ("splinter"). Relying on her knowledge of the brain, she embraces insanity as a mode of perceiving a world usually hidden behind one's normal routines:

> Much Madness is divinest Sense -
> To a discerning Eye -
> Much Sense - the starkest Madness -
> 'Tis the Majority
> In this, as all, prevail -
> Assent - and you are sane -
> Demur - you're straightway dangerous -
> And handled with a Chain - (Dickinson 1999, 278)

Besides (somewhat akin to *Sheppard Lee*) underscoring the socially constructed nature of the label "mad" ("'Tis the Majority / In this, as all, prevail -"), Dickinson here presents insanity as a way to "discern" the truth about

190   WRITING THE BRAIN

that "awful stranger - Consciousness" (1999, 512). Operating with two nuances of the word "sense" (rationality and perception), madness allows Dickinson to enter a state of "consciousness" that disrupts the "daily mind" (1999, 512), allowing her to penetrate what "mind" itself is. While an average mind in the hubbub of everyday existence discloses little about its constitution and its makeup (or, rather, what it discloses is largely invisible to a being entangled in the same society), abnormal minds disclose more. For Dickinson, a mind that stands out in full contrast to society, one that is dangerously different, clearly and visibly circumscribes its boundaries. In her poetry, facing insanity is an existential act, informing the self of itself. By analyzing, through madness, what being a self means, Dickinson points us back toward a biological force that society occasionally needs to "handle with a Chain": the brain, the implied subject that rhymes with the poem's last word.

Dickinson encountered scientific ideas about the brain and insanity through a discourse that, in a variety of ways, echoed the familiar debates of her time between dualism and monism, religion and materialism, free will and determinism. Thinking through these ideas, she found her poetic voice in the utmost expression of limitation: at the mercy of a material brain that contains all and is strictly constituted by biological laws. Faced with the reality of the brain, free will is, then, the ability not to create oneself but to confront one's own createdness, predetermined by forces outside of the control of the individual. In this utter confinement, Dickinson finds poetry—or, as she puts it in a letter mocking Matthew 6:21, "Where the treasure is, there the brain is also -" (in Habegger 2001, 509).[25] From within the brain's lobes, it appears "wider than the Sky -," "deeper than the sea -," and "just the weight of God -" (Dickinson 1999, 269). While it does not promise anything but itself—no immortality, no free will, no transcendence—the consciousness aware of its nature effuses its peculiar sense of awe:

> This Consciousness that is aware
> Of Neighbors and the Sun
> Will be the one aware of Death
> And that itself alone
>
> Is traversing the interval
> Experience between

---

[25] Matthew 6:21 (KJV): "For where your treasure is, there will your heart be also."

And most profound experiment
Appointed unto Men -

How adequate unto itself
It's [*sic*] properties shall be
Itself unto itself and None
Shall make discovery -

Adventure most unto itself
The Soul condemned to be -
Attended by a single Hound
Its own identity. (Dickinson 1999, 359)

To experience consciousness, here, is a meta-emotion of sorts. It is a constant awareness of biological limits while accepting the "Experience between" birth and death as a "profound experiment" and "Adventure." The conscious brain is indeed profound to Dickinson: it is a "deep," "penetrating" force from "below the surface" of the skull (*Webster's*, s.v. "profound"). At its core, etymologically and epistemologically, it is "adventitious" (etymologically the core of "adventure"). It is a thing "accidental," not divinely created, and "added extrinsically" through evolutionary forces (*Webster's*, s.v. "adventitious"). The brain in Dickinson's poetics hounds "It's [*sic*] own identity." The lessons Dickinson drew from her encounters with the brain through the rational, "discerning Eye" of madness place it at the center of her ongoing engagement with metaphysics, materialism, and mortality (1999, 278). When Dickinson speaks about insanity, she does so neither in a confessional mode nor attempting to self-diagnose; rather, she is performing existential experiments.[26]

Whether in an effort to satirize medicine or debate religion, Bird, Collins, and Dickinson (each in their own way) employ mind-materialism to focus on the problem of having/being an "I." Instead of merely simulating a consciousness as a first-person speaker of poem or prose, they summon "I"s that, in the extremes of being minded (and beyond), find a position to reflect on themselves as material objects. By breaking apart stable identities and "othering" the

---

[26] In this sense, Dickinson comes strikingly close to what Martin Heidegger in *Sein und Zeit* calls a state of "Sein zum Tode" (being-toward-death) that allows individuals to pull themselves out of the uniform mind of the masses ("Man") and into a state of existential authenticity ("Eigentlichkeit"). In Dickinson as in Heidegger, the conscious embrace of strict biological and temporal limits of one's life in the circumstances that one is "thrown into" (*Geworfenheit*) appears to be at the core of genius, creativity, and, ultimately, poetry (1993, 255–67).

192 WRITING THE BRAIN

conscious brain—in mad disassociation and/or brain-switching—they create a point of view from which to parse mindedness from an experiential as well as observational perspective. This "self behind the self" view allows them to reflect on the brain by oscillating between an objective and subjective perspective—authorial (*I am mind*) and possessive case (*my mind*)—and bending grammar (Dickinson) or logic (Bird) to such a degree that both positions can, at times, be inhabited simultaneously. The same holds true for the popular fascination with *The Woman in White*: emulating the character in public meant presenting a specific reading of one's mental processes—so dangerously strange—to the public. The mind, these texts and cultural metatexts underscore, is an organ that can assess the very processes that constitute it. But to achieve this feat, cognition has to be made strange, foreign. Nobody is a head-switching Captain Copp, but one can become one in writing.

Besides the philosophical quandaries that can thus be debated, Bird, Collins, and Dickinson arrest in language the paradoxical process of thinking about one's thinking. "For the unquiet heart and brain," Tennyson had commented on the relationship of writing to brain, "A use in measured language lies; / The sad mechanic exercise, / Like dull narcotics, numbing pain." Between "brain" and "pain" stands writing. As "mechanistic" as this relationship is, it can be soothing. The writing of Dickinson and Bird, in particular, was *arachnoid*, both philosophically and structurally—it saw itself as the ruff that overlaid the brain and arrested its processes in the two-dimensional space of the empty page. While such "writing as Ruff" moments were centerpieces of the works discussed here, this mode of metacognitive writing can be found in a number of texts from the early to mid-nineteenth century. We can hear hints of it in *Jane Eyre*'s confusion between mind, brain, and self in the process of thinking,[27] in *Leaves of Grass*'s pompous assertion of a text as externalized thought that can be rethought ("what I assume you shall assume"), or in the madness of the "blazing brain" that controls Captain Ahab's every decision.

"Rarely," media theorist Sybille Krämer has argued, "has a term suffered a comparable loss of reputation in recent scholarly thought as the term 'consciousness'" (1994, 88),[28] a trend she traces back to the disenchanting of "Geist" (spirit/mind) through the two-pronged attack of "naturalism" (meaning brain-centric psychology) and the computational revolution

---

[27] As the narrator frames a moment of hectic introspection: "If I *had* only a *brain active* enough to ferret out the means of attaining [a conclusion]." "I sat up in bed by way of arousing *this said brain*: it was a chilly night; I covered my shoulders with a shawl, and then *I proceeded to think*" (Brontë 1850, 84).

[28] All translations mine.

triggered by Alan Turing. Still, if we look back into the early nineteenth century, we find in the mechanistic psychology of the day a "small - italic seed" of a turning of tides.[29] In the moments of dissective metacognition described here lies the realization of the mind being *Technik*, of comprehending the conscious "I" as an operative set of natural structures and systems that produce a narrative/experience of mind. Dickinson scholar Robert Weisbuch thus rightly describes Dickinson's "strategy of writing" as an "analogical technique [that] brings us close to a mind at active work, so close that there is shocking esthetic immediacy" (1998, 212). Dickinson is practicing material cognition on the page. To Weisbuch, the poet thus becomes an inscription machine, who in the "drama of consciousness" (James again) takes us "to a mental backstage to watch the play get constructed" (1998, 214) right in front of our eyes. The brain at such moments is not a ball of tissue controlled by a godly or rationalist spark (Soul or Will) but a "brain-machine [that] is as delicate as a chronometer," as Bird himself once put it (in Foust 1919, 147). The mind is a "machinery of consciousness," as fellow doctor and Rush critic (D'Elia 1966, 227) Oliver Wendell Holmes framed it (1891–95, 264). Writing thought, to these early mind-materialists, *was* exactly the "mental technique of symbolic machines" (*Geistestechnik symbolischer Maschinen*) that Krämer finds in the (post)modern[30] split of mental processes from epistemology:

> Thus, the methodological status of the mental technologies/techniques (*Geistestechnik*) of symbolic machines [i.e., computers] has changed: What once was an accepted epistemological outlook, which had normative power to prescribe how to come to an appreciation of truth, is now refigured into a model for the functioning of mental processes as such, which matter-of-factly describes what happens when we think. The result of this reconfiguration of the epistemological methods of manipulating external symbols into an ontology of mental life is the utter loss of reputation of the category "consciousness"; it is the prime achievement of the mental techniques/technologies of symbolic machines that mental processes can be realized without a given consciousness. Where such procedures become the model for mental processes in humans, any fallback to the term "consciousness" has been rendered obsolete. (1994, 106)

[29] Of course, long before Turing, there was Charles Babbage's 1812 computing machine—so Krämer's argument about computers changing the discourse around mind could hold up if its historical reach were to be extended significantly.

[30] Krämer avoids the terminology, speaking instead of "neuzeitlich."

194   WRITING THE BRAIN

The move from "what to know" to "how to know," and the shift from normative to descriptive, is exactly what occupied figures such as Rush, Dickinson,
and Bird. In Dickinson's almost mathematical constellations of words on
the page one *can* certainly observe "mental processes . . . realized without a
given consciousness": physiological thought mimicked through the "Ruff" of
words, each thread of connection between monolithically strange metaphors
approximating the leaps and bounds of the associative engine of human
cognition.

Literarily focusing on material cognition over a disembodied soul/mind
has enabled the genesis of new forms of narrative—namely, the psychologically unreliable narrators that would become a staple of literary modernity
and postmodernity (cf. D'hoker and Martens 2008). Instead of being "unreliable" because of bias or a lack of knowledge (what Martinez and Scheffel
[2016] term "theoretically unreliable"), mind-materialists laid the groundwork for *mimetically* unreliable narrators, divorced from an objective reality
through the way their mind is structured and determined. In Dickinson's earnest experiments into the limits of cognition—and her celebration of being
at the mercy of a "machinery of consciousness"—lies the first logical step for
the fetishizing of subjectivity in the twentieth century while also heralding
its logical endpoint. Before subjectivity could become an unsurpassable gulf
between self and world, consciousness had to be decentered, brought down
to the level of muscle, soil, machinery. By realizing the limits of mind in brain
and by accepting one's inability to transcend it, these nineteenth-century
writers provide a glimpse of modernist gloom and postmodern playfulness.
The "Brain keeps giggling - still."

6

The Telegraphed Brain

Wires as Proto-Neurons

The brain is the seat of sensation, or, if you will, the central telegraph
station of the body.

—*A Handbook of the Electric Telegraph*, 1873

The whole world is the flux of matter over the wires of thought
—Ralph Waldo Emerson, *Conduct of Life*

Thoughts on Wires

The year 1803 was going to be the year of Giovanni Aldini (1734–1819).
The physics professor from Bologna, a nephew of electric mastermind Luigi
Galvani (1737–1798), had managed to acquire the right to dissect and experiment on the corpse of the notorious felon George Foster, just executed
in London. It was a chance Aldini had been waiting for. In front of members
of the Royal College of Surgeons, he applied wires to the head and body
of Foster and sent electric current through them. A newspaper account
reprinted across the United Kingdom described the results:

> On the first application of the process to the face, the jaw of the de-
> ceased criminal began to quiver, and the adjoining muscles were horribly
> contorted, and one eye actually opened. In the subsequent part of the
> process, the right hand was raised and clenched, and the legs and thighs
> were set in motion. ("England: Galvanism," 1803)

With the detail of the opening eye even famously making its way into
Frankenstein,[1] Great Britain (and, indeed, much of Europe) was quickly

[1] "By the glimmer of the half-extinguished light, I saw the dull yellow eye of the creature open; it
breathed hard, and a convulsive motion agitated its limbs" (Shelley 1823, 97–98). Another potential

Writing the Brain. Stefan Schöberlein, Oxford University Press. © Oxford University Press 2023.
DOI: 10.1093/oso/9780197693681.003.0007

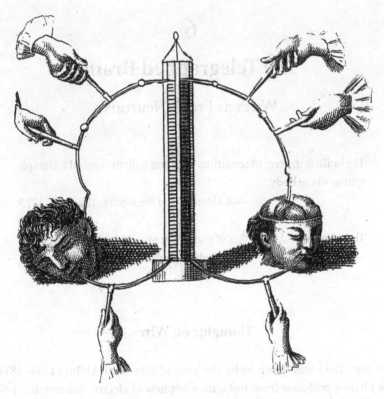

Figure 25 Aldini's experiments with brain conductors (1804, pl. 4).

abuzz with ideas of resurrecting corpses by galvanism, allowing Aldini to tour the continent to promote his name and research.

While the details of Foster's dissection are rather vague, Aldini's various books from that period go into greater detail (see Figure 25). In the years surrounding the Foster case, Aldini would take apart several different corpses and find nearly all parts of them conductive. The brain was of particular interest to him, as he conceptualized it as the central organ that "secreted animal electricity" (Butler, Rosenow, and Okun 2008, 64) and understood that "the strength and energy of the operations of the mind depend on the state of the functions of the brain" (Aldini 1803, 113).

inspiration is a similar experiment by the Scottish anatomist Andrew Ure on the corpse of the criminal Clysdale; see Siegert 2003, 276–77.

THE TELEGRAPHED BRAIN   197

In subsequent dissection cases, Aldini shocked nearly every subsection of the cerebrum he could distinguish and found them conductive to various degrees. From this research, he concluded that almost the entirety of the human body—and especially the brain—was able to transmit electricity. A dead person, he believed, may in rare cases be resuscitated by such currents, and the brain, in particular, may be shocked to cure melancholy (which he, allegedly successfully, proved by experimenting on asylum patients). Aldini also added, somewhat cryptically, that "the effects of Galvanism on the human frame differ from those produced by electricity communicated with common electrical machines" (1803, 199). Differ in kind or differ in application, he did not say.

Ultimately, Aldini's work was still vitalist, treating electricity as a mystical "life force." Beyond these claims, he reconfirmed that tissue could transmit currents and that nerves can be shocked to twitch muscle. Aldini's claims would live on in mesmerist and spiritualist circles but quickly fell out of favor with the scientific establishment once the shock from the Foster case had worn off. With his fame in steep decline, Galvani's nephew turned away from electricity. At the end of his life, he was working on lighthouses, theater illumination, and street lamps, energized exclusively by gas. Still, while the reception of his work electrifying craniums disappointed Aldini, the notion of the brain as a transmitter would return from obscurity later in the century.[2] That his outrageous claims even resulted in moderate notoriety is perhaps less a testament to Aldini than an acknowledgment of the relative dearth of robust research in the early years of electroencephalography.

This relative lack of research into the physiological makeup of brain matter changed in the 1830s with significant improvements to the lenses of compound microscopes that increased picture quality dramatically. When the German zoologist Christian Gottfried Ehrenberg (1795–1876), a friend of Humboldt's, began in that decade to look at specimens of this peculiar organ—likely using one of these cutting-edge machines[3]—he immediately

---

[2] With an eye toward Aldini, Hagner reminds us that the "possibility of bio-electricity, or magnetic currents, was only very briefly entertained by physiologists and anatomists after 1800" (2000, 119, translation mine). He later clarifies: "A reason for this [subsequent] reticence can be found in the fact that after the discovery of the voltaic pile [an early battery] around 1800, the majority of physiologists could no longer believe in animal magnetism. Experiments [on electrified tissue resulting in twitching] were now believed to be the result of . . . electricity produced directly by the contact of tissue with a fluid conductor" (2000, 188, translation mine). Volta's influential theory of contact electrification, alluded to here, was, of course, incorrect.

[3] "The most important assistance in acquiring more intimate knowledge of the phenomena of organic life is the compound microscope in the hand of the cautious practised observer" (Ehrenberg 1838, 102).

198   WRITING THE BRAIN

noticed something odd. Where most scientists of the day saw a disordered "pulpy mass, which consists of granules floating in a tough transparent fluid" (Clarke and Jacyna 1987, 61), Ehrenberg suddenly spotted *structure*. In his 1833 lecture on his results, he observed:

> The substance of the brain consists neither of granules nor of simple fibers, and is in its larger proportion formed after the model of no other texture, but consists of parallel tubules or cylindrical canals lying fasciculated upon each other . . . with remarkably regular dilations. (1838, 102)

Involved in his own philosophical battle against theories that claimed brain matter was a primordial, simple mass,[4] Ehrenberg not only confirmed some of the structural observations made by other microscopists of the day but argued that the arrangement of what he called "globules" (see Figure 26), today understood to be brain cells, was not random and that the "tubules or cylindrical canals" he observed were not fissures or accidental formations but meaningful units of matter. To him, clusters of "globules" were transmitting *something* ("nervous fluid,"[5] perhaps?)—and they were doing so via "canals."

Half a century before the first scientific description of cranial neurons by Santiago Ramón y Cajal (1852–1934), Ehrenberg's metaphors for these early hints at neuronal transmission still relied on mechanist language. To the naturalist who had studied life in the canals of Egypt (often leaving Germany via Rostock's Warnemünde canal) and lived in the "Venice of the North" (Berlin), the brain looked like a topographical map of a landscape penetrated by engineering. Hence, what was being transmitted/transported was still a passive (liquid) matter, propelled by physics, although it now moved along delicately deliberate pathways.

That very same year, the renowned physicists Carl Friedrich Gauss (1777–1855) and Wilhelm Eduard Weber (1804–1891) set up a machine just 150 miles west of Ehrenberg's home that would provide a much more electrifying set of images that could be used to talk about the brain: the electric telegraph. Indeed, Ehrenberg's observations (and the discoveries in its wake) coincide perfectly with the rise of this radically new medium—this first means of

---

[4] "Throughout his career Ehrenberg was . . . hostile to the notion that there was a primitive living *Urstoff* of which the simplest creatures were formed. . . . By contradicting that the brain was a 'polypous mass,' Ehrenberg denied a further alleged manifestation of the primitive, unorganized living substance" (Clarke and Jacyna 1987, 62).

[5] This notion is not necessarily counter to electric theory—Galvani, for instance, famously (and erroneously) believed electricity to be a fluid.

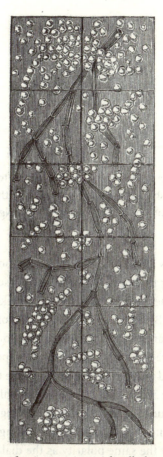

Figure 26 Human brain, after immersion in distilled water, magnified by twenty-five diameters (Ehrenberg 1838, 19).

digital transmission en masse and the first invention to allow for almost instantaneous long-distance communication. That these two media marvels of the decade, the telegraph and the brain, would soon intersect is no surprise, then, given their shared, steep rise to cultural prominence (see Figure 27). Within decades, the nineteenth century was talking wires and brains—and wired brains were only a slip of the tongue (or a spark of the mind) away. Were the body's electricity and that of "common electrical machines" (Aldini again) perhaps more closely related than hitherto believed?

What an intersection between brain metaphors and the telegraph could look like, we find amply demonstrated in the popular fiction of the day. Even before the telegraph became a public commodity and used for private

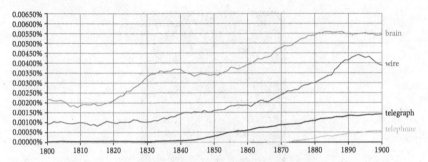

Figure 27 Relative usage of the words "brain," "wire," "telegraph," and "telephone" in the English corpus of Google Books for the years 1800 to 1900 with a smoothing value of 5 (visualized using Google Ngram Viewer).

communication, it was already fictionalized as a tool for interpersonal romance and a proxy for the mind. A particularly fascinating periodical tale titled "Valentine: Or the Electric Telegraph, A Shocking Story" came out in January 1847 in Great Britain and was thus likely composed *before* the first public telegraph company opened its doors in London (in the late summer of 1846). The brief piece by C. Sears Lancaster transposes a strictly commercial, pre-Morse railroad telegraph (the single-needle telegraph of Cooke and Wheatstone; see Figure 28) into a metaphor for the mind of a rich beer baron.

This man, subtly named Croesus Hoppy, is hunting for a perfect, highborn husband for his daughter (to acquire rank befitting his wealth) and finds that the suitors display the same polarity as the dial that translates the signals of said telegraph into words by pointing at two opposing symbols at the same time (see Figure 28). To his electrically "sympathetic mind" one suitor is "handsome and hardly-treated," the other "poor and proud" (Lancaster 1847, 125). His brain is like the dial at the center of the single-needle receiver, negotiating between them:

> Possessed of good talents, [Hoppy] entertained so absorbing an opinion of them, as to make him believe that all within his sphere were imbued with his superiority—a sort of human magnet, imparting its attractive powers to every blade that came within its contact!
>
> But—like the needle—chancing to get on the wrong side of him—the power of repulsion was in equal force. It was hit or miss with him: no wonder, then, that [the two men], wide as the poles asunder, met an exactly opposite reception. (Lancaster 1847, 135)

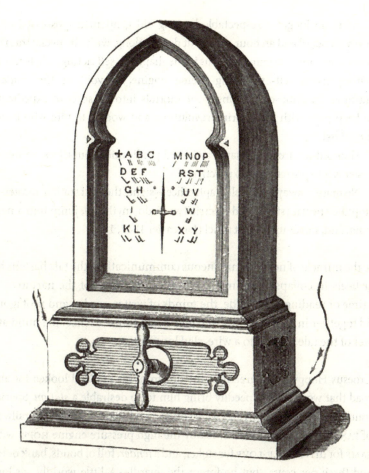

Figure 28 Cooke and Wheatstone's single-needle telegraph receiver from 1846 (*The National Encyclopædia*, pl. 2).

From this dilemma—one suitor is of bad character but of noble birth, the other his inverse—the tale takes off, overflowing with electrical and telegraphic puns, "to shew how a man got a wife by electricity (not like the eel, an electrical wife)" (Lancaster 1847, 126).

In this rather convoluted tale, the nobler of these two men proves his worthiness to marry the "useful appliance—[that] domestic automaton" (Lancaster 1847, 128) that is Hoppy's daughter—by foiling his rival's devious plot via intercepted telegraph messages. All the while, the text itself is presented as a communication akin to a rail-bound telegraph message, hurrying ever faster toward a conclusion:

202　WRITING THE BRAIN

Time is no longer a respectable-looking old gentleman, possessed of a scythe, a beard and an hour-glass: but dashes along, with his special train of woes, and comes looming upon us in the shape of a self-acting, self-feeding, self-repairing, self-everything-elseing engine, (devoid of the *tender*) startling-shocking-electrifying poor mortals into all sorts of expedients to keep-pace with his untiring revolution: and woe be to him who is *behind time*!

Humph! we are in the same awkward predicament ourselves, so much easier is it to preach than to practice.

No matter: away to the telegraph office—signal the next station-master—stop the special (steam and electricity can do anything), jump into a new train, and, make-up for lost time! (Lancaster 1847, 126)

Like the miracle of near-instantaneous communication, the tale hastens head over heels to collapse the time of narration, the time of the narrative, and the time of reading. Alongside, the minds of writer, reader, and protagonists meld together in the image of electrified brain trains. Hence, the mind at the center of the tale turns into a wire-clad locomotive:

Croesus Hoppy made greatness the dream of his life, and looked for any road that would most speedily bring him to so desirable a *station*. Special trains of thoughts and wishes were constantly dispatched upon the railway of FANCY (the swiftest extant) led by the high pressure engine HOPE; with FAME for driver, SELF-LOVE for stoker, and a *tender*, full of bonds, bank notes and three per cents., but he forgot the guard!—A little worldly wisdom perched up on the top, in the delivery of COMMON-SENSE, with instructions to sound the whistle, and use the br[ake] in the event of over-speed, might have saved the train from going "too fast"—and prevented a general "blow-up"! (Lancaster 1847, 136)

This electrified sense of self, charging full steam through one's head, is a confused metaphor, which is perhaps understandable given the infancy of its telegraphic source material. Still, it is telling. It describes a mind that is almost as swift as time itself,[6] moving on tracks (or in channels) but propelled

---

[6] Dickens, of course, would later have *his* telegraphic railroad signal actually surpass time and space and enable a communion with the dead. See Menke's excellent analysis of the "Signalman" in his *Telegraphic Realism* (2008, 166–71).

THE TELEGRAPHED BRAIN 203

by a force within, and carrying with it all qualities of mind (including the rationalist deity of "COMMON-SENSE"). Even at a time when telegraphy was still essentially a railroad signal akin to an amplified alarm bell, it seemed eerily welcoming to the mind. If thought itself, essentialized as language, could move through these curious wires clutching the rails, transmitting signals from station to station, the telegraph seemed to express a fundamental kinship with the canals and globules of the brain.

It is perhaps no surprise, then, that two years later Henry David Thoreau, upon perusing the railroad ecology around the small town of Plaistow in southeastern New Hampshire, discovered a familiar metaphor there. The wires he observed, still a somewhat rare sight with only about three thousand miles operational in the United States "at the commencement of 1848" ("The Progress of the Electric Telegraph" 1860, 290), held a strange fascination for the nature-loving transcendentalist. Otherwise generally skeptical of technological novelties, Thoreau was nonetheless mesmerized by the telegraph. Approaching the heavy iron wire, he perceived a familiar sound:

> I heard at some distance a faint music in the air like an *Aeolian harp*, which I immediately suspected to proceed from the cord of the telegraph vibrating in the just awakening morning wind, and applying my ear to one of the posts I was convinced that it was so. It was the telegraph harp singing its message through the country. . . . It was like the first lyre or shell heard on the sea-shore,—that vibrating cord high in the air over the shores of earth. (1992, 182–83)

This attempt at listening in and feeling the telegraphic vibrations would be only the first of numerous instances of Thoreau's eavesdropping on the currents of Romantic electricity that he perceives running through the wires. In moments like these, early notions of receptive minds/souls mix with concepts of minded matter and tie into broader notions of spiritualism, vitalism, and galvanism for Thoreau. In his journals, too, Thoreau records several instances of going "under the . . . wire" in the summer of 1851 (2007, 170), always leading him to press his ear against the telegraph and feel the movement of a universal, electrified order behind creation—"a spiritual telegraph permeating the universe and enabling the union of all souls," as Paul Gilmore puts it in his study of "electricity and American Romanticism" (2009, 21). The telegraph here sings of Nature, Mind, and Spirit writ large but

204   WRITING THE BRAIN

does so in the rather mundane physicality of a simple wire, used as a railroad alarm signal.

Whether Thoreau, who was aware of Ehrenberg by then (see Thoreau 1906, 240, 245), drew any connections between their telegraphic musing and early brain science is unclear, unlikely even. But that very fact is striking in and of itself: Telegraphic fictions in the vein of Thoreau and Lancaster would *anticipate* brain science by discussing notions of electric transfer of mind in material nerves half a century before it became a demonstrable fact—often decades before neuroscientists in earnest debated the idea of electro(chemical) transmission as a fundamental, structuring principle of the brain.[7] While the similarity in kind between the nerves of the body and the nervous tissue of the brain is deeply ingrained into twenty-first-century minds, in the mid-1800s that proposition was rather outrageous. Surely, the brain was a pulpy semiliquid mass, perhaps permeated by canals, not a mess of wires.

Still, in the telegraph, we see the Romantic mind metaphor of the Aeolian harp descend into a technological commodity. In this reconfigured state, said metaphor morphs into actual science. "In the eighteenth century," Wilford Spradlin and Patricia B. Porterfield have observed, "the brain was compared to clocks, cogs, gears, and pulleys. In the nineteenth century, the comparison was to be with the newly invented telegraph . . . after it was discovered that the brain utilizes electric signals" (1984, 139). That is correct—except for the term "after." Instead, the technology and its rich metaphorical echoes predated, influenced, and structured scientific discovery. With meaning made material and wired over something that so closely seemed to mimic both the nerves of the body and a common image for the mind (Aeolian harp), the material fact of mind as neuronal transmission was fictionalized as an intuited truth long before it became demonstrable fact or scientific consensus.[8]

It thus didn't take long for early neuroscience and telegraph literature to openly intersect. During the first height of telegraphic fever—the laying of

---

[7] Neither Aldini nor Ehrenberg, of course, proposed such a thing—though their theories are clearly part of a larger cultural genesis of this idea.

[8] Of course, in these years mesmerism was also proposing—from a (pseudo)scientific perspective—that electricity played a key role in cranial processes. Mesmerists went as far as stating that some parts of the brain may function akin to a Volta battery, with some proponents even calling the organ the "fountain of the nervous system" (Dods 1850, 55). The echoes of Aldini's work are obvious. Still, such concepts all developed *alongside* these telegraphic imaginaries (Mesmer himself had little to no comment on the brain) and emphasized electricity, even when stored in the brain, as a source of energy—energy that *fed* a disembodied mind and was thus categorically different from mind itself.

the transatlantic cable (1858–66)—we find the brain evoked more and more often. Fireside Poet John Greenleaf Whittier (1807–1892) in his 1858 poem "The Telegraph" (later republished as "The Cable Hymn") illustrates this exalted mood typical for telegraphic literature of the time:

> What saith the herald of the Lord?
> "The world's long strife is done!
> Close wedded by the mystic cord,
> Her continents are one.
>
> "Through Orient seas, o'er Afric's plain,
> And Asian mountains borne,
> The vigor of the Northern brain
> Shall nerve the world outworn. . . .
>
> "For lo! the fall of Ocean's wall,
> Space mocked, and time outrun!—
> And round the world, the thought of all
> Is as the thought of one!"
>
> Oh reverently and thankfully,
> The mighty wonder own!
> The deaf can hear, the blind may see,
> The work is God's alone. (1894, 256)

Overcoming the noise of the ocean, the telegraph seemed to promise an unmediated exchange of thought—physical minds blending into "one." Here, writing "about the meaning of things and the spirit of man" (Kittler 1993, 126) is not only still possible but *made possible* by the tearing down of walls between nations, between men, and between thought and language. At the height of the popular fascination with telegraphy, a wired message became "literally material thought," as a *Tribune* article informs us ("The Magnetic Telegraph" 1845, 194). Without even a hint of distortion, the telegraph brings peace through pure signification: Meaning is laid bare, essentialized to electric impulses that can literally "nerve the world" into one "brain," functioning in perfect harmony.

To Whittier, the term "mystic cord" applies to nerves as well as it does to telegraph wires. Connecting either amounts to connecting minds. It is

206   WRITING THE BRAIN

not information that is exchanged here—neither stock market prices nor messages from the British queen—but "thought" itself. Following Whittier's logic, the telegraph system mimics a system in which a "brain" uses "nerves" to transmit essential processes of mind. By glossing over the technicalities of code and the mundane mechanics of telegraph offices, maintenance workers, and system lag, Whittier and his telegraphic colleagues create a vision of direct electromechanical exchange between brains—and such transmission of mind *between* brains (already a startlingly neurofictional feast!) quickly becomes transmission within *one* brain: a world-spanning network of proto-neuronal clusters connected by wire-nerves, storing and exchanging mind by storing and exchanging electrical signals.

And indeed, science writers of the time seem to second these claims, taking up Whittier's musings and embracing the metaphor. An unsigned,[9] highly detailed 1862 piece titled "What Are the Nerves," appearing on both sides of the Atlantic (in *Cornhill Magazine* as well as the *Atlantic Monthly*), takes the telegraph system as a perfect image to comprehend the extended nervous system and the brain:

> If we look at the human brain, we find that it consists mainly of a vast mass of fibres. . . . The fibers which constitute the chief mass of the nervous system are simple in their structure, so far as the microscope can reveal it, and present a very curious analogy to a telegraphic wire. Like the latter, each nervous fiber consists of a small central thread (or tube, perhaps, in the case of the nerve . . .) surrounded by a layer of a different substance. . . . It is most natural to believe that the analogy suggested by this structure is a true one, and that the white substance [around the nerves] acts the part of the gutta percha round the electric wire, as an insulating medium for the currents which travel along the central portion. But this is not proved. (1862, 156)

Still, the unnamed lecturer abandons these notions when it comes to the brain. Instead of running with his metaphor, he adheres to notions of marvelous spirits inhabiting brain-clay; he relapses into pre-Ehrenberg beliefs, in which the brain is simply a "structure of the very lowest form," consisting of

---

[9] Since passages of the piece reappear in verbatim in Albert H. Hayes's 1878 *Diseases of the Nervous System, or, Pathology of the Nerves and Nervous Maladies*, published by the Peabody Institute, Hayes is very likely the author of the piece. Considering that Hayes served as a surgeon for the U.S. Army, his infatuation with telegraphy may have stemmed from a professional proximity to the army's many signal corps.

"mere cells and granules" that serve as a cranial tabula rasa for disembodied souls ("What Are the Nerves?" 1862, 158). The author comes ever so close to embracing the brain as a part of the nervous system—as a ball of nerves, a hub of telegraphs—but then retreats. His fellow literati of the day, however, had already ventured farther.

When it came time to celebrate the latest iteration of the transatlantic cable—its 1866 version, which worked significantly longer than previous cables—America's famous physician-poet Oliver Wendell Holmes stepped up to add a contribution with the "vigor" of *his* "Northern brain." Still, one probably had to read it twice to see its connection to the cable. It was a poem that, on its surface, celebrated a different but seemingly quite related achievement: the "fiftieth anniversary as Doctor of Medicine" of Christian Gottfried Ehrenberg. Ehrenberg had actually been involved in the construction of the cable itself; he advised its constructors on the danger of algae and microorganisms, which potentially could harm the submerged cable (Laue 1895, 213). In its crescendo, Holmes's poem honoring the microscopist exclaims:

> The smallest fibers weave the strongest bands,—
> In narrowest tubes the sovereign nerves are spun,—
> A little cord along the deep sea-sands
> Makes the live thought of severed nations one. (1868)[10]

Going as far as referencing the 1836 discovery of myelinated axons in human nerves (the "tubes" of the "nerves," akin to the protective hull of the cable) by Ehrenberg's colleague Robert Remak (1815–1865),[11] Holmes again echoes the idea of telegraphic connectivity as the world becoming one "live thought"—a thought now existing *in* nerves. To borrow a simile from Laura Otis, each "telegrapher [here] operates as an interneuron in this brain" (2011, 169)[12]—except that interneurons, or even plain neurons as building blocks of the brain, had still not yet been discovered.

---

[10] A republication of this excerpt during Holmes's lifetime (and very likely with his direct approval) confirms the connection by retitling the passage "Atlantic Telegraph, Completed, 1866" (1887, July 25–28).

[11] For more on Remak, see Wickens 2014, 165–66.

[12] While Otis's excellent treatment of "networking" as a bio-technological metaphor centers on turn-of-the-century fictions, she still at times comes close to lapsing into prochronisms when discussing the brain. The specific term "interneuron," of course, is from 1939, though the literary text it references (1898's *In the Cage*) came out a few years after the term "neuron" was first applied to the brain.

208 WRITING THE BRAIN

Telegraphed Minds

In these periodicals' engagements with the telegraph, there is something vaguely cyborgian at play, something suggesting a mind-bending interplay between human cognition and mechanical wires. This aura seemed to affect everyone connected to the project of telegraphy. The chief electrician of the transatlantic cable, for instance, certainly didn't help to dissuade such notions. Wildman Whitehouse (1816–1890) was a physician and member of the Royal College of Surgeons with a very odd personal demeanor. A *reportage* in *Chamber's Journal* of 1857 describes him as a mess of mind and wires:

> As Mr. Whitehouse stands reflectingly in his cabinet, looking at the creations of his own sagacious brain and ingenious bands, he has a cobweb of wires enveloping him, until he seems almost like some astute old spider, with its feelers out in all directions, patiently waiting for the earliest intimation of the approach of a stray visitor and victim to his net. There are wires through the ceiling, and wires through the wall, and wires through the floor; wires on this hand, and wires on that. Take heed to your steps as you venture into this suspicious-looking retreat. . . . Mr. Whitehouse has only to put now this wire and that into connection, and now to change this copper point from the binding screw to the right, so that it shall rest in the grasp of that on the left, and he is able to ask all sorts of curious and prying questions of the raw material he has to deal with; and it is very rarely indeed that he asks a question without eliciting a reply. ("The Atlantic Telegraph" 1854, 404)

If the inventor of the transatlantic cable was such a weird hybrid of man and machine, what would happen if one were to actually plug a human being into a telegraphic system? A shocking proposition—and one ready-made for periodical fiction.

In the 1858 story "An Evening with the Telegraph-Wires" by the transcendentalist Christopher Pearse Cranch, an unnamed narrator (who claims relations to a true mesmerist) recalls feeling "attracted by a low vibrating note among the leaves" while sitting in a tree (1858, 490). Finding the cause of these notes to be "three telegraph wires" (Cranch 1858, 490), he reaches out for them:

> Am I developing into *a medium*? . . . There is a telegraphic dispatch passing. Now if I could only find out what it is!—that would be something new in

THE TELEGRAPHED BRAIN   209

science,—a discovery worth knowing,—to be able to hear or feel the purport of a telegraphic message, simply by touching the wire along which it runs. So, regardless of any electric shock I might receive, I thrust out my hand through the leaves of the tree, and boldly grasped the wire. The jerks instantly were experienced in my elbow, and it was not long before certain short sentences were conveyed, magnetically, to my brain. (Cranch 1858, 490)

Literally wiretapping, the narrator not only reads the telegraphic output—business messages and private communication—but also hears them ("a mother's voice weeping over her child" [Cranch 1858, 491]), perhaps even sees them. Connected to other brains via this network, his mind shares their thoughts and impressions. In his trance-like state, these messages bring back visions to him of walking among other telegraphic networks around the world like "the long rows of telegraphic wires in France . . . looking against the sky like immense music lines" (Cranch 1858, 491). The instrument these "music lines" seem to mimic should by now be very familiar: The telegraph wires, Cranch chimes in, produce "a sigh like the first thrill of the Aeolian harp in the evening wind" (1858, 491).

In a quite similar move, the 1850s saw another acquaintance of Emerson setting out to traverse the wire. In his 1856 "Poem of Salutation," Walt Whitman's American voice poetically embraces planet Earth, noting railroads and telegraph lines, and subsuming all the impressions these technologies could bring him in an expansive poetic vision. Often lapsing into long catalogs of perceptory input, Whitman exclaims:

> Such gliding wonders! Such sights and sounds!
> Such joined unended links, each hooked to the next!
> Each answering all, each sharing the earth with all.
>
> What widens within you, Walt Whitman?
> What waves and soils exuding?
> What climes? what persons and lands are here? . . .
>
> What do you hear, Walt Whitman?
> I hear the workman singing, and the farmer's wife singing,
> I hear in the distance the sounds of children, and of animals early
>   in the day . . .

210 WRITING THE BRAIN

> What do you see, Walt Whitman? . . .
> I see the electric telegraphs of the earth,
> I see the filaments of the news of the wars, deaths, losses, gains,
> passions, of my race. (1856, 103–10)

In this internalized vision of the globe, Whitman hears and sees what lies outside of his immediate bodily grasp. Like Cranch's or Thoreau's, Whitman's senses fuse with the transmission networks of the time, nerving the world into "one" (Whittier 1894, 256). Whitman's cosmic embrace of global telegraphic communication quickly has him *experience* what it transmits. Riding the wires in a fugue of "telegraphic disembodiment" (Sconce 2000, 41), he sees "passions" and, ultimately, the whole "race" of people/brains bound together by electric wires. The wires in question have, of course, by now become organic, cranial: They have become "*filaments*" traversed by every bit of "news" that could possibly occupy a human brain as the music of the mind. Within and without, there were suddenly "joined unended links, each hooked to the next! / Each answering all"—neural connections harping their news to anyone receptive to them.

This plugging of physical, analog bodies into an electronic network of wires and nodes might strike readers as odd, but, as the narrator of "An Evening with the Telegraph-Wires" assures us, coded in the language of Hartleyian Associationism and mesmerism:

> Why should it be thought so improbable, in this age of strange phenomena, that the ideas transmitted through the electro-magnetic wire may not be communicated to the brain—especially when there exist certain abnormal or semi-abnormal conditions of that brain and its nerves? Is it not reasonable to suppose that all magnetism are one in essence? . . . is it not . . . within the scope of belief that there are those who, under certain physical conditions, may detect the purport of an electro-magnetic message,—that message being sent by vibrations of the wire through the nerves to the brain? If all magnetisms are one in essence,—as I am inclined to believe,—and if the nerves, the brain, and the mind are so swayed by what we term animal magnetism, why not allow for the strong probability of their being also, under certain conditions, equally impressible by electro-magnetism? (Cranch 1858, 493)

Often connected in popular literature to theories of "sympathy" as an organizing principle of the human body, many fiction writers poetize

THE TELEGRAPHED BRAIN   211

"electric circuitry, [as] an extension of the nervous system" (McLuhan, Fiore, and Agel 1967, 41). With language turned into signal, it was only natural for the mind to follow suit.

Emily Dickinson, a particularly "telegraphic" writer (McCormack 2003, 570), also goes all in on electromagnetic cyborgs, and presents her (imagined) readers with a romantic vision of two anthropomorphized telegraph stations in love:

> They put Us far apart -
> As separate as Sea
> And Her unsown Peninsula -
> We signified "These see" -
>
> They took away our eyes -
> They thwarted Us with Guns -
> "I see Thee" Each responded - straight
> Through Telegraphic Signs[13] -
>
> With Dungeons - They devised -
> But through their thickest skill -
> And their opaquest Adamant -
> Our Souls saw - just as well -
>
> They summoned Us to die -
> With sweet alacrity
> We stood [u]pon our stapled feet -
> Condemned - but just - to see -
>
> Permission to recant -
> Permission to forget -
> We turned our backs [u]pon the Sun
> For perjury of that -
>
> Not Either - noticed Death -
> Of Paradise - aware -

---

[13] Here Dickinson is, of course, playing with the original, pre-digital meaning of telegraphing as any "conveying information by signs to someone at a distance" (*OED*). Still, by the 1860s, this meaning had become clearly (over)charged with technological connotations.

212   WRITING THE BRAIN

> Each other's Face - was all the Disc
> Each other's setting - saw -
> (Dickinson 1999, 316–17)

Written around 1863 (and unpublished),[14] the poem seems to sing of two wired paramours, forced to talk Death, not Love, and struggling through a paradox of constant communication coupled with an inability to ever actually "see" each other.

The word "Peninsula" is telling, considering that Dickinson's *Webster's* provides this intriguing example of use: "Boston stands on a peninsula" (Webster 1828b). Boston, a crucial Civil War hub, was, of course, the location Dickinson was visiting during these months to get treatments for the mysterious ocular disease that threatened her with blindness. Her Boston doctor had advised her to reduce her writing time (he had even taken away her pen) and to stay confined in a dimly lit room with her relatives nearby. Writing to her quasi mentor T. W. Higginson in the summer of 1864, Dickinson states: "Since April in Boston for a physician's care. He does not let me go, yet I work in my prison and make guests for myself" (1906, 311).

While her letter to Higginson was composed in pencil (a formal faux pas for which she apologized), the poet likely also wired—or had somebody wire—her family back home. As Jerusha McCormack has traced, Dickinson had by then been familiar with telegraphy for almost ten years, frequently used it to keep up to date with relatives, and had specifically mentioned venturing to nearby Northampton to use its telegraph station long before her hometown was connected to the network.

The telegraph she would have seen there perhaps explains the "Disc" imagery she employs in this poem. In the Massachusetts of the 1850s and 1860s, two systems competed for dominance: the Morse telegraph and various telegraph systems that employed dials. The latter was installed in Northampton. She would have encountered telegraphy not as buzzing sounders or the scratch of a register but as discs with an indicating needle, reminiscent of a clock (see Figure 29).[15]

Telegraphy, to Dickinson, would thus be a primarily scribal, alphabetic means of communication. It was a signifying system that was itself blind but

---

[14] Given the references to Boston, this scholarly dating is likely a year off.
[15] No coincidence, since the inventor of one of these system, Alexander Bain, also invented an electric clock

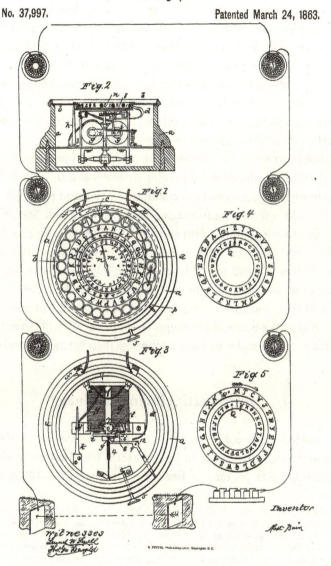

Figure 29 Patent by Alexander Bain for his 1863 "Dial Telegraph" (Google Patents).

214   WRITING THE BRAIN

could produce script—a script the eyeless lover-automatons of her poem would receive and acknowledge but could never actually read. Unlike the Morse system, which would come to be comprehended via sounders, these dial telegraphs were unintelligible without sight. Like Dickinson in her salubrious dungeon, her lover-machines are locked up in their small boxes akin to mariner's compasses ("Adamant"),[16] able only to indicate, to move a needle on the other's (clock) face. They have access to the signifier but cannot translate it into a signified. All these two receivers can share is their mutual acknowledgment of each other while their constant mental connection is being taken over by the bellicose messages of others.

Dickinson, writing with a gothic sense of humor from her own basement "prison," takes the idea of mind existing on wires at face value to summon up the ghastly image of a newborn cognition—of nascent telegraph-brains— enslaved and turned into a messaging network for others. She presents us with a tale of a telegraphic Adam and Eve,[17] "communicat[ing] ideas to another by signs" (Webster 1828c), but instead of waking in "paradise," they find themselves "stapled" to a table, locked into "Dungeon"-like casings, their electric "sweet alacrity" "summoned" to be spent on the bidding of others. These anthropomorphized telegraph devices are forced to commit "perjury"—to state the business of others—and only death will permit them to forget the horrid thoughts that moved through their extended mind network.

Introspective Brain-Machines

If there was indeed thought on the wire, the technological potential for future development seemed endless. In his wild sci-fi tale *Another World* (1873), for instance, opera aficionado Benjamin Lumley (1811–1875) depicts a

---

[16] "Mariner's compass," which in Dickinson's time referred to an object that was often called by the name of its archaic predecessor technology, the sail-stone or "adamant" (*The Chimney Corner Companion* 1827, 256), was a common way of referring to dial telegraph receivers, given their similar shapes and, perhaps, the influence of the former on inventors of the latter: "Wheatstone succeeded in finding the proper basis for arranging wires and magnets in such proportions as to produce evident effects. . . . The one upon which Cooke and Wheatstone worked was the simple fact that whenever a current of electricity passes in the neighbourhood of a magnet, such as the mariner's compass, that magnet was deflected; and Wheatstone arranged five mariner's compass needles in a horizontal row, each needle deflecting when a current of electricity was sent along the wire" (Preece 1880, 298). Given Dickinson's overt oceanic (and telegraphic) references in the poem, she seems very aware of this connection.

[17] As an 1859 telegraph manual reminds us: "The art of conveying intelligence by the aid of signals has been practised for centuries, and for aught we know since Adam and Eve commenced their pioneer career in the Garden of Eden" (Shaffner 1859, 2).

THE TELEGRAPHED BRAIN 215

telegraphically advanced society living on a different planet that also developed a quasi-magical x-ray machine to reveal the wired, electric nature of their alien brains (albeit with neither x-rays nor the electric nature of the brain yet discovered on Earth during Lumley's lifetime). In the subsection "Electrical Currents in Brain" of a chapter titled "Microscopy," the native speaker invites his readers to observe a telegraph-brain directly:

> By experiments on transparent fish of the zoophyte class, and on the eyes of animals, we discovered the means of making a living body for a time transparent. The skull was rendered transparent accordingly, and by the aid of concentrated light and of an instrument called an "electric viewer," the currents of electricity in the brain were made visible.
>
> These currents include myriads of electrical lines—literally composed of electricity—lines the nearest approach to your definition of a mathematical line, that which hath length without breadth.
>
> The filaments, as we may truly call them, are of different forms, straight, spiral, and otherwise curved, and of varied length and colours. They are set in motion by the impulsion of thought. When we talked to the patient on a particular subject, one series of lines would be set in motion with indescribable rapidity; other topics would call into play other series of straight or curved lines. They can also be set in motion under the influence of certain electricities. (Lumley 1873, 66)

Oscillating freely between the metaphorical spheres of telegraph wires, musical staff notation, and light show, Lumley has his readers briefly engage with a vision of brain activity that was clearly fantastic at the time but became actual science within decades.

With the brain hypothesized as a set of electrical lines, transmitting signals, the telegraphic literature of the latter half of the nineteenth century fostered a need for actual cranial insight. Even an end to privacy was embraced as an end to secrecy. This Cranch heralds as liberation through open data: "Instead of the tyrant hearing the secrets of the people, the people hearing the secrets of the tyrant!" (1848, 491). With little to no distinction made between encoded messages (in Morse, Wheatstone, or European letter alphabets) and thought as such, there seemed to be no obstacle to collapsing the two. So in 1854, a German music instructor walked into a patent office in London and filed to protect an invention of his that did just that. Patent no. 173 granted "Provisional Protection Only" on January 23, 1854, to Adolphus Theodore Wagner's

216 WRITING THE BRAIN

psychograph or apparatus for indicating persons' thoughts by the agency of nervous electricity (*A tracing instrument*).

The apparatus consists of a combination of rods or pieces of wood joined so as to permit of free action in all parts. From one of the legs of the instrument hangs a tracer; on one or more of the other extremities is fixed a disc, upon which the operator is to place his hand, and from this extremity or these extremities depends another tracer. The other parts of the apparatus consist of a glass slab or other non-conductor, and of an alphabet and set of figures or numerals. Upon a person possessing nervous electricity placing his hand upon one of the discs the instrument will immediately work, and the tracer will spell upon the alphabet what is passing in the operator's mind. (*Patents for Inventions* 1859, 383)

Essentially, Wagner combined an alphabetic letter grid common to early multi-needle Cooke and Wheatstone telegraphs with a set of telegraphic conductors and claimed they would let the brain write itself—with arms and fingers reduced to biological telegraph keys, transmitting "material thought" from the organ to the machine, and from there directly into signification.

Clearly, this patent was meant to describe a *transcription* machine—a secular tool, a mind-typewriter of sorts. One not too unsimilar, indeed, to Hans Berger's 1920s plans for a thought recorder that ultimately led to the first EEG (Borck 2013, 7–22). Contemporary reporting seems to suggest that Wagner's machine aimed to go beyond that and fulfill the fever dreams/nightmares of Robert Bloomfield's son to create art-automata: The machine used minds to produce poetry, created from the cranial electricity of their users, "supposedly inartistic and not especially literate Prussian citizens" (Trippett 2013, 86). A twelve-year-old girl, so it was told, was strapped into the machine and her mind produced eloquent poems celebrating Humboldt. It was the brain singing, freed from the confines of the body: pure cognition, riding the wires. The psychograph, as David Trippett states (in what appears to be the only academic analysis of the machine), may thus be "seen as an anti-metaphysical statement that embodied the *desire* to conceive of thought as a material process" (2013, 86). Still, it quickly ventured further. A Prussian officer by the name of Freiherr von Forstner immediately saw the potential of the machine and began marketing it not only in Germany and the United Kingdom but also in the United States. Within a year, psychographs were sold as far off as California (Trippett 2013, 86), and even the *New-York Daily Tribune* was covering the invention.

THE TELEGRAPHED BRAIN   217

"Nothing bears within itself a more cheering prophecy of the entire do-
minion [man] will enjoy at some future day, than does a machine or a chemical process by which labor is abbreviated and human power augmented." So
does the *Tribune* mock Wagner's original ambitions before commenting on
how marvelously his machine streamlines a specific process: table "rappings"
and like humbugs ("Help of the Ghosts" 1854). Stemming from a linguistic
sphere that uses the term "Geist" (ghost) to refer to spirits as well as minds,
and promoted in countries whose spiritualists had thoroughly embraced
telegraphic imagery (see magazines like *The Spiritual Telegraph* or *The
British Spiritual Telegraph*), Wagner's overambitious technological dream
quickly became a parlor trick. Within months, even Napoleon III had himself psychographed for his occultist enjoyment (*Napoleon the Third and His
Court* 1865, 119). By the time Wagner's far-off foreshadowing of encephalography had made it across the pond, the psychograph (then marketed as a
Seelenschreiber, a "soul writer") was already competing with an actual "spiritual telegraph" by the clockmaker Isaac T. Pease, who flat-out hooked up
spiritual mediums to a (low-voltage) telegraph machine.

Still, had musical director Wagner shopped around a bit before coming up
with his layman's idea for a mind-reading telegraph, he might have found an
actual, mechanical basis for his technophile fever dreams. A more mundane
means to measure wire signals had been around since the 1820s, but it wasn't
until 1858, when William Thomson (1824–1907) significantly improved its
sensitivity, that the galvanometer became a useful tool for detecting more
minuscule spikes in energy.

Thomson was an engineer and a board member of the Atlantic Telegraph
Company, and his contribution was perhaps the single most important step
toward realizing its ambitious plan to have Ireland and Newfoundland wire
their brains into one. The transatlantic telegraph's gigantic cables lost enough
energy on their trip that their signal was barely noticeable when it reached
the other side. But Thompson's tool was able to display even the slightest electrical transmission, by having it move a free-hanging, tiny magnet attached
to a mirror that reflected light onto a scale.[18] For this feat, he was knighted
in 1866.

[18] See also James Clerk Maxwell's 1872 Tennysonian "Lecture on Thomson's
Galvanometer: Delivered to a Single Pupil in an Alcove with Drawn Curtains": "THE lamp-light falls
on blackened walls, / And streams through narrow perforations; / The long beam trails o'er paste-
board scales, / With slow-decaying oscillations. / Flow, current! flow! set the quick light-spot flying!
/ Flow, current! answer, light-spot! flashing, quivering, dying. /¶ O look! how queer! how thin and
clear, / And thinner, clearer, sharper growing, / This gliding fire, with central wire / The fine degrees

218 WRITING THE BRAIN

When a creative Liverpool physician (and hobby Egyptologist) by the name of Richard Caton (1842–1926) got his hands on a Thomson galvanometer, he put it to more sensible (though perhaps a bit more gruesome) use than Wagner did his *Seelenschreiber*. In the early 1870s, Caton began wiring up live animals in his lab and measuring their brain activity using the telegraphic device that made the transatlantic cable speak.[19] While only a summary exists from the first public demonstration of his approach at a meeting of the Psychological Section of the British Medical Association on August 6, 1875 ("Electrical Currents of the Brain" 1875), Caton was later a frequent guest at medical gatherings in the United States,[20] where he at least once explained his approach in more detail:[21]

> I have, hitherto, experimented on the brains of forty-five animals, viz cats, rabbits, and monkeys. The instruments used were Sir William Thompson's reflecting galvanometer, with accessory apparatus [and] ... non-polarizable electrodes....
>
> The animal employed was etherized and a portion of the scalp, skull and dura mater removed, so as to expose the greater part of the hemisphere. Small, light, non-polarizable electrodes were clamped securely to the skull in such a manner that thin sculptor's clay points could be placed on any region of the exposed brain. Light insulated wires suspended from a support overhead connected the electrodes with the galvanometer. The animal

distinctly showing. / Swing, magnet! swing! advancing and receding; / Swing, magnet! answer, dearest, what's your final reading? /¶ O love! you fail to read the scale /Correct to tenths of a division; /To mirror heaven those eyes were given, / And not for methods of precision. / Break, contact! break! set the free light-spot flying! / Break, contact! rest thee, magnet! swinging, creeping, dying" (in Thompson 1910, 349).

[19] Of course, Caton learned about this approach from Helmholtz, who had used a like approach to measure nerve activity in the legs of animals. He was also inspired by the work of Fritsch and Hitzig, who were using electric shocks to the brain that resulted in the movement of animal appendages around the same time, underscoring that the human brain is *receptive* to electricity (see Hagner 2000, 274–81).

[20] During this visit, Caton, used to a nationalized and standardized telegraph system, was explicitly puzzled by the free-market chaos in the capital of the United States: "Being anxious to obtain the letters and telegrams which were addressed to us at the Congress, we made inquiry for the post and telegraph office. Unfortunately none existed in connection with the Congress, but we were referred to a neighboring post office, and were advised to visit all the telegraph offices in quest of our telegrams. As telegraphy in America is in the hands of several companies, and each company has many offices, this proved rather a formidable undertaking, as well as a fruitless one. I for one never received the telegrams sent to me in Washington" (1888, 65).

[21] On this occasion, Caton was speaking right after a doctor from Canada, who presented the curious case of a brain-damaged telegraphist who had lost 4–5 oz. of cerebral tissue but was still able to rise to a "very important office of the Western Union Telegraph Company."

THE TELEGRAPHED BRAIN   219

experimented on was tethered loosely to the center of a table a yard square, and allowed to move about, eat and drink at its pleasure. (1887, 247)[22]

His question appears simple: Does "grey matter of the brain give evidence of electrical currents comparable with those of nerve fibre and muscle"? And if so, does it correspond to function (Caton 1887, 247)? The answer was a resounding yes, a startling finding. Not only was Caton the first person to scientifically demonstrate the electric nature of the brain (and even connect it to function),[23] but he essentially developed a primitive form of electroencephalography out of telegraph technology. It would take another fifty years before Hans Berger (1873–1941), explicitly drawing on Caton's research, would publish the first human EEG.

At the time, Caton's findings attracted some interest in the transatlantic scientific community, but either because of his reluctance to publish more extensively on his experiments or because of the careful language in which he coated his findings (he stated a failure rate of 50 percent, for instance), his findings soon dropped from public consciousness. Caton even had to fight off a competitor, Adolf Beck, for the achievement of having discovered the electric nature of the brain more than a decade later. By the turn of the century, Caton had turned away from brain research and toward local politics (he became lord mayor of Liverpool) as well as the study of ancient Egypt—a much more scientifically developed culture than archeologists gave it credit for, he emphasized (1904, 19).

While Caton's research largely receded into popular obscurity, the telegraph-brain nexus did not. Although functioning mammal brains had a lot to say about their wired nature, it was a dysfunctional brain-mind that perhaps made the most eloquent use of it. Lloyd G. Thompson's "The Zig Zag Telegraph; A Tale of a Mind Diseased," a slightly belated entry into the genre (published in 1884, but likely written much earlier),[24] still relies on a mere pen to record its

---

[22] The test was likely inspired by Emil du Bois-Reymond's famous 1848 "Twitch Telegraph"—a galvanometer used to measure electrical activity in arm and hand muscles via conducting plates (somewhat akin to Scientology's "E-Meter"). The Twitch Telegraph, of course, was too blunt an instrument to measure the minute electricity of the brain. For more on the tool, see Finkelstein 2015.

[23] "Fluctuations of the current were observed sometimes, coinciding with some movement of the animal's body or some change in its mental condition. In deep anæsthesia, for example, considerable deflection of [the] needle was generally observed, lessening gradually as the animal regained consciousness" (Caton 1887, 247).

[24] The story begins by prefacing that its writer had waited for "nearly nineteen years" to see it into print (Thompson 1884, 184), while the rest of the tale lacks this timely remove from its contents. The concluding paragraph, too, suggests that a draft of "The Zig Zag Telegraph" would have been written in the months, perhaps years, following Thompson's release from captivity, but certainly not decades later. The piece ends on a hopeful note and is optimistic about the protagonist's recovery—an

220 WRITING THE BRAIN

eponymous mind. Thompson's autobiographical tale, published in the *Atlantic*, fuses budding notions of introspection as a psychological technique with the technophile imagery of preceding decades to describe a case of what would now be called PTSD that is startlingly modern. It also provides a firsthand account of experiencing oneself as an electro-telegraphic cognition. "The Zig Zag Telegraph" is the author's personal narrative of suffering from an odd mental condition as a result of his Civil War service, a condition doctors at the time would have described as an extreme form of homesickness—a "nostalgia" that afflicted only less hardy soldiers (Benedek and Wynn 2011, 4; see also Horwitz 2015). Thompson did not subscribe to such notions. Instead, he makes the case, in an at times surprisingly humorous tone, that his abnormal state of mind has nothing to do with weakness or a longing for home; rather, he suffers from a case of a *rewired brain* as a result of trauma.

His readers at the time must have been puzzled by Thompson's eloquent report. Some reviewers were so confused they misread the tale as "a contribution to the scientific mind mysteries of to-day, such as clairvoyance, mind-readings, etc." ("Book Notices" 1884). Startlingly, the latter concepts are completely absent from the short piece. Perhaps the reviewer had given up and filled in his or her memory gaps with some Twainesque technomysticism. But even reviewers who did find in Thompson's account "an interesting study of the unconscious workings of the brain" ("The Magazines for August" 1884, 53) or a detailed description of an "overstrained brain" ("[Reviews]" 1884, 76) lacked a specific language (like "trauma" or "PTSD") to truly comprehend what the author was describing: None of the short literary notices about the tale even mentions war.

Still, Thompson, a seldom-published writer who has eluded scholarship so far, is very forthcoming about his experiences during the conflict and his months in confinement in the notorious prisoner-of-war camp at Andersonville.[25] What his tale sets out to achieve is to describe the mental state resulting from a truly "exceptional condition of body or mind" (1884, 184) that

optimism that, after almost two decades, would have been either confirmed or proven wrong. The introductory paragraph thus read like they were composed at a later date then the rest of the piece.

[25] A biographical sketch of Lloyd Garrison Thompson (1837–1904) is found in a Thompson family genealogy: "He enlisted in the war of the rebellion. August, 1863, in Company F, 76th New York volunteer infantry. Was in the army of the Potomac over two years . . . He was taken prisoner May 5, 1864, in the battle of the Wilderness; was imprisoned at Danville, Andersonville, and Florence; paroled March 6, 1865, a mere skeleton, a physical wreck . . . After the war he again lived in Orleans county [New York]; later Fair Haven, Conn. Is a writer of merit" (Thompson 1890, 172). John Worrell Northrop's 1904 *Chronicles from the Diary of a War Prisoner in Andersonville* contains detailed

THE TELEGRAPHED BRAIN 221

has afflicted Thompson ever since his camp experiences; and he hopes to do so by "materializ[ing] this action" of the brain (1884, 186) via the well-traveled image of the telegraph. What was to follow was "a system of exploration in the realms of memory" (1884, 184) relying on wires, stations, and signals.

To give Thompson's less attentive reviewers credit, his opening paragraphs do tease a tale more in line with occultist "Mental Telegraphy" (as Twain termed telepathy) than a biographical account of the long-term effects of wartime violence:

> For nearly nineteen years I have been waiting for some one to write the history of this line; but during all this time no account of its origin, or the manner in which it performed its work, has been published, and so far as I can learn no hint even of its existence has appeared in print. Can it be possible that I was sole proprietor and operator; that my weary messages alone went creeping over the wires; that its faithful, patient services were given to me only? If so, upon me clearly devolves the task of writing its history. And yet, to own the truth, this task is not an easy one. The Zig Zag was such an anomaly among telegraphs, such a *bizarre* affair altogether, that it sets at defiance all ordinary methods of description. (1884, 184)

How enticing! Still, if one holds on to learn about all the fabulous, otherworldly workings of this mysteriously teased "Zig Zag,"[26] one is in for a disappointment. Instead of metaphysical marvels, Thompson presents expressly *material* detours, dead ends, and delays: physiological, mental processes outside of the control of this brain-telegraph's proprietor.

What Thompson is describing would today be called "flashbacks": glimpses of past trauma. Once triggered, these flashbacks interfere with normal mental activity. "Flashback," of course, is a filmic term unavailable to Thompson. To him, the telegraph must do. "I will represent memory as a network of

accounts of Thompson, who apparently read Milton to fellow prisoners and knew Longfellow poems by heart. Upon first meeting him, Northrop noticed that a "mellow beam of genial nature shone in his face" (Northrop 1904, 59), and he expressed admiration for Thompson's attempts at aiding wounded soldiers in the camp. Besides letters to the editor on a number of issues (including on prisoner's pension bills and the "horrors of Andersonville") and a few poems (like his Civil War poem "The Boys Across the River") printed in local newspapers, no other works by Thompson could be located. "The Zig Zag Telegraph" seems to have been his only national publication.

[26] As a term, it is a not altogether innocent one. Its (German) origins in warfare resurfaced in midcentury military jargon: "Zig-zag is not only the proper course by which to advance in sieges, but it is the method of connecting the parallels and places of arms, and finally arriving at the close of the attack or breaching batteries" (*Aide-Mémoire to the Military Sciences* 1862, 792).

222 WRITING THE BRAIN

telegraph wires," Thompson clarifies, "the main line connecting the mind with the beginning of conscious existence, and side wires connecting this line with each event, each incident, each thought, of past life." Before he can explain his particular condition, he lays out his wired model of the mind:

> When the mind is unimpaired and the lines are in perfect working order, information can be obtained instantly from any of these out-lying stations. The question is flashed over the wires, and the answer is returned, and the combined messages constitute a thought. In many instances, however, no perceptible action of the mind seems required; the mind is unquestioning and at rest; and yet, from the various depots in which our experiences of the past are stored, the messages come trooping in, and we call them memories. These are phases of the normal action of the intellect and the undisturbed working of the lines. But I am also familiar with many phases of abnormal action, and various stages of wreck in the lines of communication. (Thompson 1884, 186)

He then continues to describe full and partial amnesia as "the wires all down" or "the main line cut at a given point," respectively, and memory loss in age as "side lines . . . mostly broken and impaired" except for those "farther back" (Thompson 1884, 186). Thompson, who had served under the telegraphically adept Major General William Starke Rosecrans,[27] had himself experienced such metaphorical "downed lines" in his immediate experience of camp fever and starvation, but his lasting effects are different still. These constitute not cut lines but a whole different set of lines altogether, zigzagging through his mind and connecting his main line of thought with unwanted memories and feelings.

This "Zig Zag" line Thompson demonstrates using a number of vivid examples of miswired mental connections that cause seemingly random lapses into memory during an interview for his medical furlough papers before he "[breaks] down utterly, and [cries] like a child" (1884, 189):

> I gave the name of my regiment, and the officer dashed it down, and asked brusquely, "What company?" I ought to have been prepared for this question, but I was not. My mind was so dazed with the strange workings of the two lines that I thought of nothing else till the question was put. Again I turned to the past, and inquired, "What company?" and again the Zig Zag took the question.

---

[27] Rosecrans played a major role in standardizing and reorganizing the Army Signal and Telegraph Service and relied heavily on wiretapping.

First message by Zig Zag. A river spanned by a bridge; beyond the bridge an arch of evergreens and flags; a throng of men hurrying over the bridge and under the arch; the men are emaciated and half naked, but their faces glow with joy.

Second message by Zig Zag. A forest; Union soldiers grouped round a dead cavalry man; a sergeant with face turned toward the group, as if about to give an order; a line of Confederate troops in front.

By the Direct Line, Company F.

I named the company, and the officer jotted it down, and said, "Your captain's name?" Again the Zig Zag took the question.

First message by Zig Zag. A long line of Union soldiers, with a group of officers on horseback in front; the officers with field-glasses to their eyes; the ground in front descending to a small stream, then ascending to a ridge; the ridge crowned with a line of Confederate earth-works and batteries; sharpshooters deployed as skirmishers between the lines.

Second message by Zig Zag. A prison pen; a scaffold; six men with ropes around their necks and meal sacks drawn over their heads; a sea of faces turned up toward the scaffold.

Third message by Zig Zag. A brigade drawn up in hollow square; a man kneeling on a coffin, with a file of soldiers before him; an officer standing stern and pale, his extended right hand holding a white handkerchief.

By the Direct Line, Captain Goddard.

The exchange goes on like this. These unwanted "messages" to Thompson are accompanied by sweats, light-headedness, and panic as he fears his "mind [may be] hopelessly shattered" (1884, 188).

Thompson's is not only a startlingly detailed (and rare, especially for the nineteenth century) first-person account of the experience of PTSD that just happens to find itself expressed through a common metaphor for the brain, but also a clear example of said metaphor actively structuring the way Thompson comprehends his condition. Not only does his vision of main lines (thick, fully developed, direct) and support lines (thinner, slower, decaying with neglect) anticipate scientific theory on the varying strength and number of different neural connections in the brain, but Thompson also becomes obsessed with *transmission speed*:

If A, B, and C represent stations on the Zig Zag, and D the desired point on the Direct Line, and the interval of time between messages was five seconds, my

224 WRITING THE BRAIN

> message would be five seconds in reaching A, and the return message from A would reach me five seconds later, or at the exact time that my dispatch reached B; while the message from B would reach me at the same instant that my dispatch reached C, and consequently the message from C would reach me at the same time that my dispatch reached D, the point on the Direct Line. (1884, 190)

Certainly, this is a point where Thompson may have lost some of his reviewers. But also in this moment, we fully observe the brain becoming a medium: In transmission, it overcomes space using time, and in memory, it overcomes time by arresting it in space. Thought is suddenly a physical signal that needs time to traverse its cranial distances. Now capable of time axis manipulation (Krämer 2006), the brain shares not only the *potential* of technology but also its downsides: It experiences network lag, faulty wires, and short circuits.

Even when the electrochemical nature of the brain was still a mere cautious proposition (certainly, Thompson would have been unaware of Caton), the telegraph allowed minds like Thompson's to comprehend themselves as a cognitive medium: a thing that, in its very structure, shared a fundamental similarity with machines. In tales like Thompson's, there can no longer be a transcendent mind inhabiting a simple, primordial mass, pulling on nerve strings to connect the outside world. Now there was just a system of links between memory banks, with thought turned into electrical transmission and memory into data storage.

Relying on the cerebral aura of the telegraph and his own experience as an enmattered mind alone, Thompson could—more than a century before modern brain imaging would confirm it—describe a case of PTSD as something that had long-term effects on the material nature of the mind. The war, Thompson could telegraphically argue, fundamentally changed the cranial pathways through which his mind found expression, as well as the intensity with which its constituent "stations" functioned. Long before neuroscientists would be able to demonstrate "smaller hippocampal and anterior cingulate volumes, increased amygdala function, and decreased medial prefrontal/anterior cingulate function" and show "increased cortisol and norepinephrine responses to stress" in PTSD patients (Bremner 2006, 445), Thompson had developed an account of the mind startlingly congruous with such findings.

N. Katherine Hayles has famously described a "feedback loop" between "materiality and discourse," between technology and culture (1993, 153). The nexus between brain science and telegraph culture is perhaps one of the best examples of such a loop, with the barriers between imaginative proposition and techno-scientific discovery blurring to such a degree that a clear-cut differentiation between each is, while perhaps not impossible, clearly counterproductive. Drawing from the harp-brain loop of earlier decades, the mid-1800s saw microscopists' discoveries structured by their cultural notions of transmission media, authors of sci-fi and techno-satire proposing the wiring of brains to machines before actual scientists took them up on it, and communication technology offering a blueprint to early neuroscience that was both metaphorical and physiological.

The final outcome of this loop is the neuron doctrine—the fundamental idea that the human nervous system includes the brain, which is likewise made up of distinct cells in electro(chemical) communication with one another, and that said cells specialize. While the actual doctrine was only formulated in the early twentieth century, it divulges cultural-technological midwifery of the mid- to late nineteenth century. Perhaps nobody embodies this insight better than Santiago Ramón y Cajal,[28] the father of the neuron doctrine, whose famous drawings of the minute structure of the brain exude such an eerie sense of beauty that they are now exhibited in art museums. But as otherworldly as his 1905 illustrations seem, to a technologically inclined reader their electric filaments may look familiar (Figure 30):

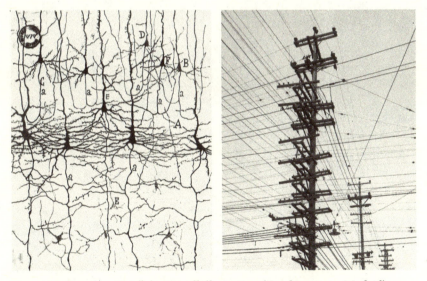

Figure 30 Ramón y Cajal's brain-cell illustration (Cajal Institute, Madrid) vs. urban telegraph wires (Fremont Historical Society).

---

[28] Ramón y Cajal, too, relied on the telegraph when discussing the brain (Friedrich 2012, 143).

Afterword

"Mathematics, music, and poetry," the son of poet-harpist Robert Bloomfield had exclaimed in 1844, "are all composed of order and novelty, or of symmetry and variety, which are either blended or alternately exhibited" (300). Automata that produce art were thus only a matter of time. The insights gleaned from hooking brains up to telegraphic galvanometers—the neuronal doctrine—may ultimately usher this dream/nightmare of the young watchmaker into reality. A glance at the following metacognitive poem from 2022 may serve as an apt illustration. It takes us back, albeit in a more modern tone, to the poem by Dr. Abraham Coles that introduced this book. It reads:

> The brain is a wondrous thing,
> A marvel of evolution's reign,
> A universe of neurons firing,
> A symphony of electrical wiring.
>
> It's the seat of our consciousness,
> The source of our thoughts and emotions,
> It controls our every move,
> And holds the key to all we do.
>
> From the simplest tasks to the most complex,
> The brain is there to guide us,
> It's the master of our movements,
> And the keeper of our memories.
>
> With its endless capacity to learn and grow,
> The brain is truly one of a kind,
> A unique and wondrous creation,
> A gift that we should cherish and mind.
>
> So let us celebrate the brain,
> This most miraculous of organs,

228   AFTERWORD

> And give thanks for all it does,
> For without it, we would be lost.

A straightforward and rather clumsy piece, certainly, but not an altogether inartistic set of verses, considering that it was, in fact, composed by a computer algorithm named ChatGPT. A conceptual possibility since the 1940s, and computationally a reality since the 2010s, generative machines like these (in this case, a third-generation "generative pre-trained transformer") consist of "neural circuits in which neurons are abstracted into nodes and linked by connections that change through learning rules" (Yuste 2015, 490). ChatGPT, in large part, is an artificial neural network, a computer-learning algorithm that consumes large corpora of sources (be they textual, visual, or auditory) and learns to recognize patterns and, in turn, produce them (while never just regurgitating source material verbatim).

ChatGPT isn't even a poetry specialist but a chatbot that mimics human expression. In this instance, it created the above poem in under a second in response to the prompt "Write me a poem about the brain." It even seems to pun, suggesting we should "cherish and *mind*" the brain. The result is a fascinating bundle of paradoxes: ChatGPT is a piece of software developed by approximating human cognitive processes (first theorized in the 1800s by comparing human brains to machines), which is now creating art about the complexity of mindedness.

ChatGPT, like other artificial neural networks, does not merely execute preprogrammed commands (if x then y) but "learns": The more it "reads," the better it performs. Approximating biological neurons, its decision nodes (termed "artificial neurons") activate in clusters, but these clusters are not predetermined by a human but rather are generated by the machine based on the input it has digested. In many ways, ChatGPT is the perfect, belated endpoint to this story of the first century of the brain: a realization of the most technophile and far-fetched fantasies the 1800s could muster, which are now, finally, catching up with the infrastructure necessary to turn them into reality.

Between 1800 and 1880, the mind descended from being a (quasi)immaterial entity into the form of pulpy lobes and folds of the brain, whose structure enabled and limited the thinking self that was generated in and from this matter. To think about thinking in these years meant considering biological structures, be they cranial hemispheres, the topography of the brain, specific structures within it, or the electronic wires that, surely, must run

AFTERWORD   229

through it. Writers of science and literature alike participated in this debate, hypothesizing and illustrating concepts of material cognition, often in the very same venues. Specific movements in (proto)neuroscience found kindred minds within budding literary (sub)genres: The dual mind was itself as Gothic as the best fiction of Poe or Melville; phrenology stretched itself into neurosociology with the help of the mid-Victorian Realism of Whitman and Dickens (ultimately lapsing into psychometrics); insanity became a metacognitive trope in Collins and Dickinson; and the structuring metaphors of Aeolian harp and telegraph suggested mental processes to be the result of wired transmission.

In many ways, *Writing the Brain* has been a recovery effort. It aimed to excavate the literary echoes, collaborations, and interrogations of anatomically justified theories of mind. These theories were beginning to emerge in the early 1800s and began receding from public discourse toward the end of the century with the emergence of more clearly delineated, modern scientific disciplines and a larger split of brain science from popular culture. *Writing the Brain* not only demonstrated that elements of this new mind-materialism were present in canonical texts on both sides of the Atlantic but also introduced a number of obscure and understudied minor texts into this conversation, hoping to roughly chart a transatlantic cranial corpus that enables further exploration by literary and (other) media scholars.

To more clearly outline the physical brain in the texts of its first rise to prominence, it was necessary to largely divorce this analysis, at least heuristically, from the emergence of a related mental science: psychology. Especially the story of Victorian psychological theories morphing into a modern, scientific discipline has generated a wealth of important scholarship. Its bracketing off here has been not a dismissal of these narratives but a necessary disentangling of scientific epistemologies. The pitfalls and promises of material monism were at the heart of *Writing the Brain*; psychology, especially in its early day, almost inevitably required a degree of dualism to make a case for its efficacy. Of course, rarely are early psychology and the "proto-neuroscience" of this book perfectly distinct, either in individual authors or in genres. A book like Tyson Stole's *Dickens and Victorian Psychology*, for instance, reminds us that Dickens could "insist upon the immateriality and immortality of mind" (2022, 241) while showing an "openness to phrenology" that feels "unexpected" and runs counter to these convictions (2022, 76). To make sense of the "unexpected," it was necessary to read outside of the dominant narrative of the emergence of modern psychology.

230 AFTERWORD

The initial fever pitch of enthusiasm for the newly mapped-out brain that was at the heart of this book soon turned sour when it failed to produce the spectacular outcomes it seemed to promise, be it in the form of novel technologies, medical treatments, or societal reform. Popular interest waned, and more and more scientists (such as Caton and Ramón y Cajal) chose to promote their findings exclusively in medical journals, resulting in a reduced cultural footprint of the brain in the decades to follow. Still, since the closing decades of the twentieth century, neuromateriality has returned to popular culture in full force (from transhumanist sci-fi and the neuronovel to the work of Google's "Brain Team"). The questions Dickinson asked about struggling material minds are perhaps even more pressing in the age of neuropharmacology, and Whitman's phrenological musings about absorbing cognitive difference into democratic society could be read as an early insistence on inclusion and neurodiversity.

In a time when a chatbot like ChatGPT can proclaim our collective platitudes about the mysteries of mindedness back to us, a rereading of the dawn of material cognition may be a welcome antidote to cultural amnesia. "[The brain] controls our every move, / And holds the key to all we do," this neural network tells us, emphasizing that our various material cognitions (biological and technological) constitute a shared existential quandary. And while the questions pondered by Brontë and Bird, Collins and Coleridge may ultimately end in realizing their unresolvability, they nonetheless add much-needed depth and context—to present debates about artificial cognition and material minds, as well as to nineteenth-century scholarship. In its poem, ChatGPT extols the power of this "most miraculous of organs," still echoing the fundamental confusion and awe of Dr. Coles for this "weird, and wonderful, and fragile mass" (1866, 38). Machines and humans alike, it seems, have yet to fully grapple with the nature of cognition.

"The metamorphization of the human body, mind, and soul," Jonathan Sterne has rightly noted, "follows the medium currently in vogue" (2003, 289). This book has suggested that such a process of metamorphization is neither unidirectional nor straightforward. Brain, body, and mind are as much media themselves as they attract technological similes. The brain is, then, perhaps the ultimate "material substrate of culture," as it fuses interior (as experience) as well as exterior (as stuff, material) conditions of semantics (Siegert 2015, 3). What has been commonly understood as a process of ever-shifting, trendy metaphors is much more complex than that. The brain is not

AFTERWORD   231

just comparable, for instance, to a telegraph—but it is an epistemic object in and of itself, created as culture and disclosing culture.

"The guiding question" for theoretical inquiry in this vein, Siegert observes with a nod toward cybernetics, is "not 'How did we become posthuman?' but rather, 'How was the human always already historically mixed with the nonhuman?'" (2015, 6). The problem of "artificial intelligence" is then not merely a question for sci-fi and fantasy. Instead "intelligence," in its material expression as traced here, is always already "artificial:" It is an "art," a *cultural technique*, following Macho, with the "ability to thematize [itself]" (2013, 30).

In various ways, then, *Writing the Brain* told a set of counterhistories about how we frame what Foucault termed the "figure of Man" and Kittler polemicized as "so-called Man." The former famously described this fiction of subjectivity as born of a rationalist science built on sand and situated close to the method of its own erasure. There is a sneaking disregard for nineteenth-century minds encoded in such observations—they imply a past that wanted to *know* about the self but was unaware of the contradictions and potential erasures of self encoded into that impulse. *Writing the Brain*, I hope, has provided a few notes indicating that this was not the case. As these preceding chapters have demonstrated, the various cerebral exchanges between scientific cultures and literary cultures were acutely aware of the epistemological chaos their endeavors caused.

In place of Foucault's face in the sand, perhaps Emerson's "Circles" may be a more apt image for the conclusion of the present volume—another ode to the end of "so-called Man" and perhaps a more fitting celebration of its demise as a logical endpoint of the drive to understanding subjectivity and mind:

> There is no outside, no inclosing wall, no circumference to us. The man finishes his story,—how good! how final! how it puts a new face on all things! He fills the sky. Lo! on the other side rises also a man, and draws a circle around the circle we had just pronounced the outline of the sphere. Then already is our first speaker not man, but only a first speaker. His only redress is forthwith to draw a circle outside of his antagonist. And so men do by themselves. The result of to-day, which haunts the mind and cannot be escaped, will presently be abridged into a word, and the principle that seemed to explain nature will itself be included as one example of a bolder generalization. (Emerson 2012b, 189)

232 AFTERWORD

To Emerson, cognition is a circle composed of concepts produced by other circles that bloom into "mere" signifiers in death. As Christopher Hanlon has demonstrated (2018), this impulse has always been part of Emerson, from early pieces like "Circles" all the way to his late style, marked by Emerson's own mind refracting through a brain experiencing dementia. In his work, not only is the Concord Sage only fully conscious of the grammatology of knowledge production, but he implies that the further one digs into the concept of mind, the more it appears multitudinous, atomic, contradictory. Cognition is not a state but a process. It is less *mind* than *thinking*: material strategies executed by and within matter. To a Coleridgean cognition like Emerson's, that is cause for celebration—his decentering of Man is joyous, not a Foucauldian burial at sea.

Material, cranial introspection, as traced through the discourses in this book, is not just the determinist specter that critics often hope to exorcise from analyses of its technological haunts. The experimental outlook that guided Dickinson's pen into materializing and externalizing thought into script or that allowed Thompson to frame his subjectivity as relay networks is, in a very real sense, related to current debates on AI art, neural network journalism, and computers that can visually recreate thoughts. This relationship is not merely one of *Gedankengeschichte* (history of thought) but one birthed by the very contradictions encoded into the material condition of the mind. The textual techniques discussed in this volume underscore that our worries about the limits and exploitability of material minds merely constitute a return of deferred questions. The fact that human subjects, overall, do not behave as autarkic, self-centered subjects has by now metastasized into consumer psychology, addiction-centered screen interaction, and neuropsychological design. The thrill, perhaps even joy, that can be found in the realization of one's cerebral multitudes and contradictory cognitions remains largely unrealized.

References

"Accident." 1825. *Gettysburg Compiler*. May 18, 1825: 4.

Ackerman, Michael E. 2009. "Phantoms of Old Forms: The Gothic Mode in the Dramatic Verse of Tennyson and Browning." PhD dissertation, Wilfrid Laurier University.

Agg, John. 1819. *The Ocean Harp: A Poem; in Two Cantos*. Philadelphia: M. Thomas.

Aide-Mémoire to the Military Sciences. 1862. London: John Weale, Lockwood & Co.

Aldini, Giovanni. 1803. *An Account of the Late Improvements in Galvanism: With a Series of Curious and Interesting Experiments*. London: Cuthell & Martin and J. Murray.

Aldini, Giovanni. 1804. *Essai théorique et expérimental sur le galvanisme*. Paris: De l'Imprimerie de Fournier fils.

Almond, Philip C. 1982. *Mystical Experience and Religious Doctrine: An Investigation of the Study of Mysticism in World Religions*. Berlin: Mouton.

Alter, Jonathan. 2013. "Obama's 'Double Consciousness' on Race." *The New Yorker*, July 25, 2013. www.newyorker.com/news/news-desk/obamas-double-consciousness-on-race.

Althubaiti, Turki S. 2015. "Race Discourse in 'Wuthering Heights.'" *European Scientific Journal* 11 (8): 201–25.

"Amusements – Meetings – Lectures." 1874. *Springfield Republican*. February 3, 1874: 1.

Anderson, Charles R. 1959. "The Conscious Self in Emily Dickinson's Poetry." *American Literature* 31 (3): 290–308.

Annual Report of the Trustees of the Northampton Lunatic Hospital for the Year Ending Sept 30 1877. 1877. Boston: Rand, Avery & Co.

Anogianakis, George. 2014. "Reflections of Western Thinking on Nineteenth Century Ottoman Thought: A Critique of the 'Hard-Problem' by Spyridon Mavrogenis, a Nineteenth Century Physiologist." In *Brain, Mind and Consciousness in the History of Neuroscience*. Edited by C. U. M. Smith and Harry A. Whitaker, 91–104. Dordrecht: Springer.

Aspiz, Harold. 1966. "Educating the Kosmos: 'There Was a Child Went Forth.'" *American Quarterly* 18, no. 4: 655–66.

"The Atlantic Telegraph." 1854. *Chamber's Journal of Popular Literature*, no. 182: 401–4.

Auerbach, Erich. 1953. *Mimesis: The Representation of Reality in Western Literature*. Princeton, NJ: Princeton University Press.

Austen, Jane. 1899. *Persuasion*. Boston: Little, Brown and Co.

Austen, Jane. 1906. *Persuasion: With Colored Illustrations by C. E. and H. E. Brock*. New York: Frank S. Holby.

Axelrod, Alan. 1983. *Charles Brockden Brown: An American Tale*. Austin: University of Texas Press.

Babbage, Charles. 1864. *Passages from the Life of a Philosopher*. London: Longman, Green, Roberts & Green.

Bain, Alexander. 1873. *Mind and Body: The Theories of Their Relation*. London: H. S. King.

234 REFERENCES

Bardi, Abby. 2008. "'Gypsies' and Property in British Literature: 'Orlando' and 'Wuthering Heights.'" In *"Gypsies" in European Literature and Culture*. Edited by Valentina Glajar and Domnica Radulescu, 105–23. New York: Palgrave Macmillan.

Barker, Fred G., II 1995. "Phineas Among the Phrenologists: The American Crowbar Case and Nineteenth-Century Theories of Cerebral Localization." *Journal of Neurosurgery* 82 (4): 672–82.

Barnard, Philip, and Stephen Shapiro. 2006. "Introduction." In *Edgar Huntly: Or, Memoirs of a Sleep-Walker*. Edited by Philip Barnard and Stephen Shapiro, ix–xlii. Indianapolis: Hackett.

Barrish, Phillip. 2005. *White Liberal Identity, Literary Pedagogy, and Classic Realism.* Columbus: The Ohio State University Press.

Baumgartner, Barbara. 2016. "Anatomy Lessons: Emily Dickinson's Brain Poems." *Legacy* 33 (1): 55–81.

Beard, John Relly. 1853. *The Life of Toussaint L'ouverture, the Negro Patriot of Hayti, Etc.* London: Ingram.

Beecher, Henry Ward. 1891. *Patriotic Addresses in America and England from 1850 to 1885: On Slavery, the Civil War and the Development of Civil Liberty in the United States.* New York: Fords, Howard & Hulbert.

Bell, Charles. 1802. *The Anatomy of the Brain: Explained in a Series of Engravings*. London: T. N. Longman and O. Rees.

Bell, John, and Charles Bell. 1803. *The Anatomy of the Human Body: Containing the Nervous Systems with Plates*, vol. 3. London: T. N. Longman and O. Rees.

Bell, Richard. 2012. *We Shall Be No More: Suicide and Self-Government in the Newly United States*. Cambridge, MA: Harvard University Press.

Bender, Bert. 1988. *Sea-Brothers: The Tradition of American Sea Fiction from Moby-Dick to the Present*. Philadelphia: University of Pennsylvania Press.

Benedek, David M., and Gary H. Wynn. 2011. *Clinical Manual for Management of PTSD.* Washington, DC: American Psychiatric Publications.

Benn, Gottfried. 1910. "Beitrag Zur Geschichte Der Psychiatrie." *Die Grenzboten: Zeitschrift für Politik, Literatur und Kunst* 69: 92–95.

Bennett, Jane. 2010. *Vibrant Matter: A Political Ecology of Things*. Durham, NC: Duke University Press.

Bersani, Leo. 1976. *A Future for Astyanax: Character and Desire in Literature*. Boston: Little, Brown.

Bethune-Baker, James Franklin. 1901. *The Meaning of Homoousios in the "Constantinopolitan" Creed*: Cambridge University Press.

Bird, Robert Montgomery. 1841. *The Difficulties of Medical Science: An Inaugural Lecture, Introductory to a Course of Lectures*. Philadelphia: Pennsylvania Medical College.

Bird, Robert Montgomery. 2008. *Sheppard Lee: Written by Himself*. Edited by Christopher Looby. New York: New York Review of Books.

Bischoff, Theodor Ludwig Wilhelm. 1874. *Ueber den Einfluss des Freiherrn Justus von Liebig auf die Entwicklung der Physiologie: Eine Denkschrift von Dr. Theodor L. W. Von Bischoff*. Munich: Verlag der k.b. Akademie.

Blanckaert, Claude. 2011. *De la race à l'évolution: Paul Broca et l'anthropologie française (1850–1900)*. Paris: Harmattan.

Bloomfield, Robert. 1824. "Nature's Music: Consisting of Extracts from Several Authors: With Practical Observations, and Poetical Testimonies, in Honour of the

REFERENCES 235

Harp of Æolus." In *The Remains of Robert Bloomfield*, 1:95–186. London: Baldwin, Cradock, and Joy.

Bloomfield, Robert Henry. 1844. "The Mechanics of Music." *The Musical World* 19: 293–294; 300–301.

Boehm, Katharina. 2013. *Charles Dickens and the Sciences of Childhood: Popular Medicine, Child Health and Victorian Culture*. New York: Palgrave Macmillan.

Boelhower, William. 1987. *Through a Glass Darkly: Ethnic Semiosis in American Literature*. New York: Oxford University Press.

Bondevik, Hilde. 2010. "Who's Afraid of Amalie Skram? Hysteria and Rebellion in Amalie Skram's Novels of Mental Hospitals." In *Illness in Context*. Edited by Knut Stene-Johansen and Frederik Tygstrup, 181–98. Leiden: Brill.

"Book Notices." 1884. *The Index* 5 (5): 58.

Boos, Sonja. 2021. *The Emergence of Neuroscience and the German Novel: Poetics of the Brain*. Cham, Switzerland: Springer International.

Borck, Cornelius. 2013. *Hirnströme: Eine Kulturgeschichte Der Elektroenzephalographie*. Göttingen: Wallstein Verlag.

Borst, Anton. 2014. "A Chant of Dilation: Walt Whitman, Phrenology, and the Language of the Mind." PhD dissertation, City University of New York.

Boshears, Rhonda, and Harry Whitaker. 2013. "Phrenology and Physiognomy in Victorian Literature." In *Literature, Neurology, and Neuroscience: Historical and Literary Connections*. Edited by Anne Stiles, Stanley Finger, and François Boller, 87–112. Amsterdam: Elsevier.

Bourgery, Jean-Baptiste Marc, and Nicolas Henri Jacob. 1844. *Traité complet de l'anatomie de l'homme*, vol. 3. Paris: C.-A. Delaunay.

Boziwick, George. 2014. "'My Business Is to Sing': Emily Dickinson's Musical Borrowings." *Musical Women in Nineteenth-Century America* 8 (2): 130–66.

Brande, William Thomas. 1853. *A Dictionary of Science, Literature, and Art: Comprising the History, Description, and Scientific Principles of Every Branch of Human Knowledge. With the Derivation and Definition of All the Terms in General Use*. New York: Harper & Brothers.

Brasher, Thomas L. 1958. "Whitman's Conversion to Phrenology." *Walt Whitman Newsletter* 4: 95–97.

Bray, Charles. 1871. *A Manual of Anthropology: Or, Science of Man, Based on Modern Research*. London: Longmans & Co.

Bremner, J. Douglas. 2006. "Traumatic Stress: Effects on the Brain." *Dialogues in Clinical Neuroscience* 8 (4): 445–61.

Brewer, William M. 1930. "Poor Whites and Negroes in the South Since the Civil War." *Journal of Negro History* 15 (1): 26–37.

"The British Association and the Negro." 1863. *The Leeds Mercury* September 2, 1863: 2.

Brock, William Hodson. 1967. *The Atomic Debates: Brodie and the Rejection of the Atomic Theory; Three Studies*. Leicester: Leicester University Press.

Brontë, Charlotte. 1850. *Jane Eyre: An Autobiography*. London: Smith.

Brontë, Emily. 1848. *Wuthering Heights*. New York: Harper & Brothers.

Brown, Charles Brockden. 1798. "Queries." *The Weekly Magazine of Original Essays, Fugitive Pieces, and Interesting Intelligence*, June 23, 1798. Charles Brockden Brown Archive.

Brown, Charles Brockden. 2006. *Edgar Huntly: Or, Memoirs of a Sleep-Walker*. Edited by Philip Barnard and Stephen Shapiro. Indianapolis: Hackett.

236 REFERENCES

Brown, Thomas. 1827. *A Treatise on the Philosophy of the Human Mind: Being the Lectures of the Late Thomas Brown, M.D.; Abridged, and Distributed According to the Natural Divisions of the Subject*, vol. 1. Boston: Hilliard and Brown.

Brown, William Wells. 1855. *The American Fugitive in Europe: Sketches of Places and People Abroad*. Boston: J. P. Jewett.

Brown, William Wells. 1863. *The Black Man: His Antecedents, His Genius, and His Achievements*. Boston: James Redpath.

Brown, William Wells. 1864. *Clotelle: A Tale of the Southern States*. Boston: J. Redpath.

Brown, William Wells. 1874. *The Rising Son: Or, the Antecedents and Advancement of the Colored Race*. Boston: A. G. Brown.

Bruce, Dickson D., Jr. 1992. "W. E. B. Du Bois and the Idea of Double Consciousness." *American Literature* 64 (2): 299–309.

Brunström, Conrad. 2004. *William Cowper: Religion, Satire, Society*. Lewisburg, PA: Bucknell University Press.

Burdge, Julia. 2016. *Introductory Chemistry: An Atoms First Approach*. New York: McGraw-Hill.

Burrell, Brian. 2003. "The Strange Fate of Whitman's Brain." *Walt Whitman Quarterly Review* 20 (3): 107–33.

Burrell, Brian. 2004. *Postcards from the Brain Museum: The Improbable Search for Meaning in the Matter of Famous Minds*. New York: Broadway Books.

Butler, Matthew A., Joshua M. Rosenow, and Michael S. Okun. 2008. "History of the Therapeutic Use of Electricity on the Brain and the Development of Deep Brain Stimulation." In *Deep Brain Stimulation in Neurological and Psychiatric Disorders*. Edited by Daniel Tarsy, Jerrold L. Vitek, Philip A. Starr, and Michael S. Okun, 63–82. Totowa, NJ: Humana Press.

Byerly, Alison. 1997. *Realism, Representation, and the Arts in Nineteenth-Century Literature*. Cambridge: Cambridge University Press.

Cahan, David. 1993. *Hermann Von Helmholtz and the Foundations of Nineteenth-Century Science*. Berkeley: University of California Press.

"Cambridge, April 4th." 1811. *Norfolk Chronicle and Norwich Gazette*. April 6, 1811: 3.

Carlson, Eric T., Jeffrey L. Wollock, and Patricia S. Noel. 1981. "Earlier Theories." In *Benjamin Rush's Lectures on the Mind*. Edited by Eric T. Carlson, Jeffrey L. Wollock, and Patricia S. Noel, 392–402. Philadelphia: American Philosophical Society.

Carlyle, Thomas. 1840. *On Heroes, Hero-Worship and the Heroic in History*. London: Chapman and Hall.

Carter, Boyd. 1953. "Poe's Debt to Charles Brockden Brown." *Prairie Schooner* 27 (2): 190–96.

Castonguay, Stephane. 1996. "Un iatrochimiste du Bas-Canada: Francois Blanchet et les Recherches sur la medecine ou application de la chimie a la medecine." *Bulletin canadien d'histoire de la medécine* 13 (2): 315–31.

Caton, Richard. 1887. "Researches on Electrical Phenomena of Cerebral Gray Matter." *Transactions of the International Medical Congress* 3: 246–50.

Caton, Richard. 1888. "Notes of a Visit to the Ninth International Medical Congress." *Liverpool Medico-Chirurgical Journal* 8 (14): 64–78.

Caton, Richard. 1904. *I-Em-Hotep and Ancient Egyptian Medicine: The Harveian Oration Delivered Before the Royal College of Physicians on June 21, 1904, Etc*. London: C. J. Clay & Sons.

Cattell, J. 1846. "On the Non-Duality of the Mind." *Lancet*, February 7, 1846: 151–52.

REFERENCES 237

Chaouli, Michel. 2002. *The Laboratory of Poetry: Chemistry and Poetics in the Work of Friedrich Schlegel*. Baltimore: Johns Hopkins University Press.

Chelius, J. M. 1847. *A System of Surgery*. Edited by George W. Norris. Philadelphia: Lea & Blanchard.

Cheshire, Paul. 2001. "The Eolian Harp." *Coleridge Bulletin* 17: 1–22.

Child, Lydia Maria. 1833. *An Appeal in Favor of That Class of Americans Called Africans*. Boston: Allen and Ticknor.

The Chimney Corner Companion. 1827. London: Knight & Lacey.

Chinoy, Sahil. 2019. "The Racist History Behind Facial Recognition." *New York Times*. July 10, 2019. https://nytimes.com/2019/07/10/opinion/facial-recognition-race.html.

"City News." 1874. *Massachusetts Spy*. February 13, 1874: 1.

Claeys, Gregory. 2014. "'The Only Man of Nature That Ever Appeared in the World': 'Walking' John Stewart and the Trajectories of Social Radicalism, 1790–1822." *Journal of British Studies* 53 (3): 636–59.

Claggett, Shalyn. 2011. "George Eliot's Interrogation of Physiological Future Knowledge." *SEL Studies in English Literature 1500–1900* 51 (4): 849–64.

Clarke, Basil. 1987. *Arthur Wigan and the Duality of the Mind*. Psychological Medicine Monograph Supplement 11. Cambridge: Cambridge University Press.

Clarke, Edwin, and L. S. Jacyna. 1987. *Nineteenth-Century Origins of Neuroscientific Concepts*. Berkeley: University of California Press.

Cleman, John. 1991. "Irresistible Impulses: Edgar Allan Poe and the Insanity Defense." *American Literature* 63 (4): 623–40.

Cochran, Peter. 2014. *The Farmer's Boy by Robert Bloomfield: A Parallel Text Edition*. Newcastle upon Tyne: Cambridge Scholars Publishing.

Coffman, Peter W. 2015. "'He Has Earned the Right of Citizenship': The Black Soldiers of North Carolina in the Civil War. A Comment on Historiography, Treatment, and Pensions." Master's thesis, East Carolina University.

Coleridge, Samuel Taylor. 1848. *Hints Towards the Formation of a More Comprehensive Theory of Life*. Philadelphia: Lea and Blanchard.

Coleridge, Samuel Taylor. 1884. *Complete Works of Samuel Taylor Coleridge: With an Introductory Essay upon His Philosophical and Theological Opinions*. Edited by W. G. T. Shedd. New York: Harper & Brothers.

Coleridge, Samuel Taylor. 1895. *Letters of Samuel Taylor Coleridge*. Edited by Ernest H. Coleridge. Boston: Houghton, Mifflin.

Coleridge, Samuel Taylor. 1912. *The Poems of Samuel Taylor Coleridge: Including Poems and Versions of Poems Now Published for the First Time*. Edited by Ernest H. Coleridge. London: Oxford University Press.

Coleridge, Samuel Taylor. 1959. *Collected Letters: 1815–1819*. Edited by Earl L. Griggs. Oxford: Clarendon Press.

Coleridge, Samuel Taylor. 1992. *Collected Works of Samuel Taylor Coleridge: Marginalia*. Edited by H. J. Jackson, Kathleen Coburn, and Bart K. Winer. London: Routledge.

Coleridge, Samuel Taylor. 2002. *Coleridge's Notebooks: A Selection*. Edited by Seamus Perry. Oxford: Oxford University Press.

Coles, Abraham. 1866. *The Microcosm: A Poem, Read Before the Medical Society of New Jersey at Its Centenary Anniversary; with the Address Delivered as President, Etc.* New York: D. Appleton & Company.

The College Board. 2016. "The SAT: Practice Test 7." http://www.launchpadeducation. com/wp-content/uploads/2017/10/SATTest7w_coversheet.pdf.

238 REFERENCES

Collins, Wilkie. 1868. *The Moonstone: A Novel.* New York: Harper & Brothers.

Collins, Wilkie. 2013. *Woman in White.* Mineola, NY: Dover Publications.

Combe, George. 1830. *Essays on Phrenology.* London: John Anderson.

Combe, George. 1836. *The Constitution of Man: Considered in Relation to External Objects.* Edinburgh: Maclachlan, Stewart, and John Anderson.

Combe, George. 1841. *The Constitution of Man: Considered in Relation to External Objects.* Boston: Marsh, Capen, Lyon, and Webb.

Combe, George. 1860. *A System of Phrenology.* New York: Harper & Brothers.

Conrad, Lawrence I. 1995. *The Western Medical Tradition: 800 BC to AD 1800.* Cambridge: Cambridge University Press.

"The Conjoint St Leger." 1861. *Bell's Life in London and Sporting Chronicle.* October 20, 1861: 8.

Cooke, John. 1824. *A Treatise on Nervous Diseases.* Boston: Wells and Lilly.

Coole, Diana, and Samantha Frost, eds. 2010. *New Materialisms: Ontology, Agency, and Politics.* Durham, NC: Duke University Press.

Cooper, B. B. 1832. *Lectures on Anatomy: Interspersed with Practical Remarks.* 4 vols. London: Longman, Reese, Brown, Green.

Crain, Caleb. 2001. *American Sympathy: Men, Friendship, and Literature in the New Nation.* New Haven, CT: Yale University Press.

Cranch, Christopher P. 1858. "An Evening with the Telegraph-Wires." *Atlantic Monthly* 2 (11): 489–95.

Cromwell, Thomas. 1859. *The Soul and the Future Life. The Philosophic Argument.* London: Edward T. Whitfield.

Cunningham, Andrew, and Nicholas Jardine, eds. 1990. *Romanticism and the Sciences.* Cambridge: Cambridge University Press.

"Curious Epigram on Philip Syng Physick." 1859. *Medical and Surgical Reporter* 3 (4): 110.

Cutter, Calvin. 1847. *Anatomy and Physiology: Designed for Academies and Families.* Boston: Benjamin B. Mussey.

Dalton, John. 1808. *A New System of Chemical Philosophy.* Manchester: S. Russell.

Dalton, Lisle. 2002. "Phrenology and Religion in Antebellum America and Whitman's *Leaves of Grass.*" *Mickle Street Review* (15): 1–36.

Damasio, H., T. Grabowski, R. Frank, A. M. Galaburda, and A. R. Damasio. 1994. "The Return of Phineas Gage: Clues About the Brain from the Skull of a Famous Patient." *Science* 264 (5162): 1102–105.

Danby, John Francis. 2000. "William Wordsworth: Poetry, Chemistry, Nature." In *The Green Studies Reader: From Romanticism to Ecocriticism.* Edited by Laurence Coupe, 44–49. London: Routledge.

"Dangers of Sleepwalking." 1824. *The Raleigh Register.* July 16, 1824: 3.

Daniels, Melissa Asher. 2013. "The Limits of Literary Realism: Of One Blood's Post-Racial Fantasy by Pauline Hopkins." *Callaloo* 36 (1): 158–77.

Darwin, Charles. 1871. *The Descent of Man, and Selection in Relation to Sex: In Two Volumes* 1. London: John Murray.

Davis, Andrew S., and Raymond S. Dean. 2005. "Lateralization of Cerebral Functions and Hemispheric Specialization: Linking Behavior, Structure, and Neuroimaging." In *Handbook of School Neuropsychology.* Edited by Rik C. D'Amato, Elaine Fletcher-Janzen, and Cecil R. Reynolds, 120–44. Hoboken, NJ: Wiley.

Davy, Humphry. 2011. *Memoirs of the Life of Sir Humphry Davy.* Edited by John Davy. Cambridge: Cambridge University Press.

REFERENCES   239

Davy, John. 1839. *The Collected Works of Sir Humphry Davy: Early Miscellaneous Papers from 1799 to 1805, with an Introductory Lecture and Outlines of Lectures on Chemistry, Delivered in 1802 and 1804.* London: Smith, Elder.

Dawson, Terence. 1989. "The Struggle for Deliverance from the Father: The Structural Principle of 'Wuthering Heights.'" *Modern Language Review* 84 (2): 289–304.

De Forest, John William. 1867. *Miss Ravenel's Conversion from Secession to Loyalty.* New York: Harper & Brothers.

de Quincey, Thomas. 1886. *Confessions of an English Opium-Eater.* London: W. Scott Publishing Company.

de Quincey, Thomas. 1876. *The Works of Thomas De Quincey,* vol. 3. London: Hurd and Houghton.

Deakin, Wayne George. 2015. *Hegel and the English Romantic Tradition.* London: Palgrave Macmillan.

D'Elia, Donald J. 1966. "Dr. Benjamin Rush and the American Medical Revolution." *Proceedings of the American Philosophical Society* 110 (4): 227–34.

Descartes, René. 1687. *L'homme de René Descartes, et La formation du foetus; ou Traité de la lumière du mesme autheur.* Paris: Girard.

Descartes, René. 2012. *Principles of Philosophy: Translated, with Explanatory Notes.* Edited by Valentine R. Miller and Reese P. Miller. Dordrecht: Springer Netherlands.

Desmond, Adrian J., and James R. Moore. 2009. *Darwin's Sacred Cause: How a Hatred of Slavery Shaped Darwin's Views on Human Evolution.* Boston: Houghton Mifflin Harcourt.

D'hoker, Elke, and Gunther Martens, eds. 2008. *Narrative Unreliability in the Twentieth-Century First-Person Novel.* Berlin: de Gruyter.

Dickens, Charles. 1881. *Great Expectations.* Boston: Estes and Laureat.

Dickens, Charles. 1933. *Charles Dickens's Letters to Charles Lever.* Edited by Flora V. M. Livingston. Cambridge, MA: Harvard University Press.

Dickens, Charles. 1971. *Our Mutual Friend.* Edited by Stephen Gill. Harmondsworth: Penguin.

Dickinson, Edward. 1821. "The Importance of Providing an Asylum." The Houghton Library, Harvard University.

Dickinson, Edward. 1865. "Letter to Pliny Earle." American Antiquarian Society.

Dickinson, Edward. 1870. "Letter to Pliny Earle." American Antiquarian Society.

Dickinson, Emily. 1906. *The Letters of Emily Dickinson 1845–1886.* Edited by Mabel L. Todd. Boston: Little, Brown.

Dickinson, Emily. 1971. *The Life and Letters of Emily Dickinson.* Edited by M. D. Bianchi. New York: Biblo and Tannen.

Dickinson, Emily. 1999. *The Poems of Emily Dickinson.* Edited by R. W. Franklin. Cambridge, MA: Belknap Press.

Dictionary of American Family Names. 2003. Edited by Patrick Hanks. Oxford: Oxford University Press.

Diderot, Denis, and Jean Rond Le d'Alembert. 2017. *Encyclopédie, ou dictionnaire raisonné des sciences, des arts et des métiers, etc.* Edited by Robert Morrissey and Glenn Roe. Chicago: ARTFL Encyclopédie Project, University of Chicago.

Dix, Robin C. 1988. "The Harps of Memnon and Aeolus: A Study in the Propagation of an Error." *Modern Philology* 85 (3): 288–93.

Dods, John Bovee. 1850. *The Philosophy of Electrical Psychology: In a Course of Twelve Lectures.* New York: Fowler and Wells.

240 REFERENCES

Downing, Harriet. 1837. "Remembrances of a Monthly Nurse." *Fraser's Magazine for Town and Country* 16: 497–511.

"A Dual Brain." 1853. *Southern Standard.* March 12, 1853: 3.

Dunant, M. 1875. "Description of M. Kastner's New Musical Instrument, the Phyrophone." *Journal of the Society of the Arts* 23 (1161): 293–96.

Dybikowski, James. 2008. "Joseph Priestley, Metaphysician and Philosopher." In *Joseph Priestley, Scientist, Philosopher, and Theologian.* Edited by Isabel Rivers and David L. Wykes, 80–112. Oxford: Oxford University Press.

Earle, Pliny. 1854. *An Examination of the Practice of Bloodletting in Mental Disorders.* New York: Wood.

Earle, Pliny. 1864. "Letter to Edward Dickinson." John Hay Library, Brown University.

Earle, Pliny. 1898. *Memoirs of Pliny Earle, M.D.* Edited by F. B. Sanborn. Boston: Damrell & Upham.

Edwards, Jonathan. 1840. *The Works of Jonathan Edwards.* Edited by Edward Hickman. London: Ball, Arnold and Co.

Edwards, Paul. 1967. *The Encyclopedia of Philosophy*, vol. 7. New York: Macmillan.

Ehrenberg, Christian Gottfried. 1838. "Observations on the Structure Hitherto Unknown of the Nervous System in Man and Animals." In *Essays on Physiology and Hygiene.* Edited by John Bell, 67–120. Philadelphia: Haswell, Barrington and Haswell.

"Electrical Currents of the Brain." 1875. *Chicago Journal of Nervous & Mental Disease* 2 (4): 610.

Eliot, George. 1856. "The Natural History of German Life." *Westminster Review*, July 1856: 28–43.

Eliot, George. 1859. *Adam Bede.* New York: Harper & Brothers.

Eliot, George. 2003. *Daniel Deronda.* Ware, MA: Wordsworth Classics.

Elliotson, John. 1835. *Human Physiology.* London: Longman, Orme, Brown, Green, and Longmans.

Ely, Ezra Stiles. 1822. *A Synopsis of Didactic Theology.* Philadelphia: J. Crissy.

Emerson, Ralph Waldo. 1849. *Nature.* Boston: James Munroe.

Emerson, Ralph Waldo. 2001. *The Later Lectures of Ralph Waldo Emerson: 1843–1871.* Edited by Ronald A. Bosco and Joel Myerson, vol. 2. Athens, GA: University of Georgia Press.

Emerson, Ralph Waldo. 2012a. *Emerson: Poems.* New York: Knopf Doubleday.

Emerson, Ralph Waldo. 2012b. *The Annotated Emerson.* Cambridge, MA: Belknap Press.

"England: Galvanism." 1803. *Cork Mercantile Chronicle.* January 28, 1803: 4.

Erhardt-Siebold, Erika von. 1931–1932. "Some Inventions of the Pre-Romantic Period and Their Influence upon Literature." *Englische Studien* 66: 347–63.

Ewen, Elizabeth, and Stuart Ewen. 2008. *Typecasting: On the Arts and Sciences of Human Inequality, a History of Dominant Ideas and Stuart Ewen.* New York: Seven Stories.

"Extraordinary Accident and Cure." 1850. *The New York Times.* December 6, 1850: 3.

Fairer, David. 2009. *Organising Poetry: The Coleridge Circle, 1790–1798.* Oxford: Oxford University Press.

Farr, Judith. 2005. *The Gardens of Emily Dickinson.* Cambridge, MA: Harvard University Press.

Feltenstein, Rosalie. 1947. "Melville's 'Benito Cereno.'" *American Literature* 19 (3): 245–55.

Fiedler, Leslie A. 1997. *Love and Death in the American Novel.* Normal, IL: Dalkey Archive Press.

REFERENCES 241

Finger, Stanley. 1994. *Origins of Neuroscience: A History of Explorations into Brain Function*. Oxford: Oxford University Press.

Finkelstein, Gabriel. 2015. "Mechanical Neuroscience: Emil Du Bois-Reymond's Innovations in Theory and Practice." *Frontiers in Systems Neuroscience* 9 (113): https://doi.org/10.3389/fnsys.2015.00133.

Finney, Carolyn. 2014. *Black Faces, White Spaces: Reimagining the Relationship of African Americans to the Great Outdoors*. Chapel Hill: University of North Carolina Press.

Fisher, Benjamin F. 2002. "Poe and the Gothic Tradition." In *The Cambridge Companion to Edgar Allan Poe*. Edited by Kevin J. Hayes, 72–91. Cambridge: Cambridge University Press.

Fisher, Philip. 1999. *Still the New World: American Literature in a Culture of Creative Destruction*. Cambridge, MA: Harvard University Press.

"Five Hundred Dollar Reward." 1831. *Christian Advocate and Journal and Zion's Herald*. October 7, 1831. https://americanantiquarian.org/NatTurner/items/show/39.

Fleischman, John. 2010. *Phineas Gage: A Gruesome but True Story About Brain Science*. Dubuque, IA: Kendall/Hunt.

Fletcher, A. K. 2015. "Racial Typology and the Pseudosciences in Harriet Beecher Stowe's *Uncle Tom's Cabin*." https://scrollsofakfletcher.wordpress.com/2015/12/30/racial-typology-and-the-pseudosciences-in-harriet-beecher-stowes-uncle-toms-cabin/.

Ford, Brian J. 2007. "Enlightening Neuroscience: Microscopes and Microscopy in the Eighteenth Century." In *Brain, Mind, and Medicine: Essays in Eighteenth-Century Neuroscience*. Edited by Harry A. Whitaker, C. U. M. Smith, and Stanley Finger, 29–43. New York: Springer.

Fornal, Justin. 2016. "Exclusive: Inside the Quest to Return Nat Turner's Skull to His Family." https://nationalgeographic.com/news/2016/10/nat-turner-skull-slave-rebellion-uprising/.

Foust, Clement E. 1919. *The Life and Dramatic Works of Robert Montgomery Bird*. New York: Knickerbocker Press.

Fowler, Lorenzo, and Orson Squire Fowler. 1849. "Republicanism, and Its Improvements." *The American Phrenological Journal and Miscellany* 11: 208–15.

Fowler, Orson Squire. 1847. *Physiology, Animal and Mental: Applied to the Preservation and Restoration of Health of Body, and Power of Mind*. New York: Fowler and Wells.

Fowler, Orson Squire, and Lorenzo Fowler. 1836. *Phrenology Proved, Illustrated and Applied*. New York: W. H. Coyler.

Friedrich, Alexander. 2012. "Metaphorical Anastomoses: The Concept of 'Network' and Its Origins in the Nineteenth Century." In *Travelling Concepts for the Study of Culture*. Edited by Birgit Neumann, Ansgar Nünning, and Mirjam Horn, 119–44. Berlin, Boston: de Gruyter.

Gall, Franz Joseph, and Johann Caspar Spurzheim. 1815. *Outlines of the Physiognomical System of Drs. Gall and Spurzheim: Indicating the Dispositions and Manifestations of the Mind*. London: Baldwin, Cradock, and Joy.

García-Molina, A. 2002. "Phineas Gage and the Enigma of the Prefrontal Cortex." *Neurología* 27 (6): 370–75.

Gaskell, Elizabeth C. 1900. *The Life of Charlotte Brontë*. New York: Harper & Brothers.

Gilbert, Sandra M., and Susan Gubar. 2000. *The Madwoman in the Attic: The Woman Writer and the Nineteenth-Century Literary Imagination*. New Haven, CT: Yale University Press.

242 REFERENCES

Gilman, Sander L. 1986. "Black Bodies, White Bodies: Toward an Iconography of Female Sexuality in Late Nineteenth-Century Art, Medicine, and Literature." In *"Race," Writing, and Difference*. Edited by Henry Louis Gates Jr. and Kwame A. Appiah, 223–61. Chicago: University of Chicago Press.

Gilmore, Paul. 2009. *Aesthetic Materialism: Electricity and American Romanticism*. Stanford, CA: Stanford University Press.

Glass, Jesse. 2006. *The Passion of Phineas Gage and Selected Poems*. Sheffield, UK: Westhouse Books.

Goodridge, John. 2002. "Robert Bloomfield and the Aeolian Harp: A Conversation with Alan & Nina Grove." *Robert Bloomfield Society Newsletter*, March, 7–13.

Gordon, Lyndall. 2014. *Lives Like Loaded Guns: Emily Dickinson and Her Family's Feuds*. New York: Penguin Books.

Gorsky, Susan R. 1999. "'I'll Cry Myself Sick': Illness in Wuthering Heights." *Literature and Medicine* 18 (2): 173–91.

Gosline, Sheldon Lee. 2014. "'I Am a Fool': Dr. Henry Cattell's Private Confession About What Happened to Whitman's Brain." *Walt Whitman Quarterly Review* 31 (4): 158–62.

Gould, Stephen Jay. 2008, 1996. *The Mismeasure of Man*. New York: W. W. Norton.

Gower, Barry. 1973. "Speculation in Physics: The History and Practice of Naturphilosophie." *Studies in History and Philosophy of Science* 3 (4): 301–56.

Grant, A. Cameron. 1965. "Combe on Phrenology and Free Will: A Note on XIXth-Century Secularism." *Journal of the History of Ideas* 26 (1): 141–47.

Gravil, Richard. 2015. *Wordsworth's Bardic Vocation, 1787–1842*. London: Palgrave, Macmillan.

Greding, John Ernst. 1798. "Medical Aphorisms on Melancholy: And Various Other Diseases Connected with It." In *An Inquiry into the Nature and Origin of Mental Derangement: Comprehending a Concise System of the Physiology and Pathology of the Human Mind. And a History of the Passions and Their Effects*. Edited by Alexander Crichton, 2:349–445. London: Cadell Jr. and W. Davies.

Greenberg, Kenneth S. 2004. *Nat Turner: A Slave Rebellion in History and Memory*. Oxford: Oxford University Press.

Griffin, John Joseph. 1849. *Chemical Recreations: A Popular Compendium of Experimental Chemistry, for the Use of Beginners*. London: John J. Griffin.

Griffin, John Joseph. 1860. *The Chemistry of the Non-Metallic Elements and Their Compounds: Air—Water—the Gases—the Acids*. London: John J. Griffin and R. Griffin.

Gross, Theodore L. 1961. "The Negro in the Literature of Reconstruction." *Phylon* 22 (1): 5–14.

Grovier, Kelly. 2005. "Shades of the Prison-House: 'Walking' Stewart, Michel Foucault and the Making of Wordsworth's 'Two Consciousnesses.'" *Studies in Romanticism* 44 (3): 341–66.

Grovier, Kelly. 2007. "Dream Walker: A Wordsworth Mystery Solved." *Romanticism* 13 (2): 156–63.

Habegger, Alfred. 2001. *My Wars Are Laid Away in Books: The Life of Emily Dickinson*. New York: Random House.

Hagner, Michael. 2000. *Homo Cerebralis: Der Wandel vom Seelenorgan zum Gehirn*. Frankfurt am Main: Insel Verlag.

Haller, John S. 1996. *Outcasts from Evolution: Scientific Attitudes of Racial Inferiority, 1859–1900*. Carbondale: Southern Illinois University Press.

REFERENCES 243

Hamilton, Cynthia S. 2008. "'Am I Not a Man and a Brother?': Phrenology and Anti-Slavery." *Slavery & Abolition* 29 (2): 173–87.

Hanes, Susan R. 2015. *Wilkie Collins's American Tour, 1873–4.* London: Routledge.

Hanlon, Christopher. 2018. *Emerson's Memory Loss: Originality, Communality, and the Late Style.* Oxford: Oxford University Press.

Harlow, John M. 1869. *Recovery from the Passage of an Iron Bar Through the Head.* Boston: Clapp.

Harper, Douglas. n.d. *Online Etymology Dictionary.* Accessed February 23, 2017. http://etymonline.com.

Harrington, Anne. 1987a. *Medicine, Mind, and the Double Brain: A Study in Nineteenth-Century Thought.* Princeton, NJ: Princeton University Press.

Harrington, Anne. 1987b. "[Review of Psychological Medicine Monograph Supplement 11]." *Psychological Medicine* 19 (1): 245–63.

Harte, Richard H. 1906. "Philip Syng Physick: A Sketch." *Medical Bulletin* 18 (12): 339–44.

Hartley, David. 1810. *Observations on Man: His Frame, His Duty and His Expectations. To Which Are Added, Prayers, and Religious Meditations.* London: Wilkie and Robinson.

Hayes, Kevin J. 2013. *Edgar Allan Poe in Context.* Cambridge: Cambridge University Press.

Hayles, N. Katherine. 1993. "The Materiality of Informatics." *Configurations* 1 (1): 147–70.

Hayles, N. Katherine. 2010. *How We Became Posthuman: Virtual Bodies in Cybernetics, Literature, and Informatics.* Chicago: University of Chicago Press.

Hedges, William. 1974. "Charles Brockden Brown and the Culture of Contradictions." *Early American Literature* 9 (2): 107–42.

Heidegger, Martin. 1993. *Sein Und Zeit.* Tübingen: Max Niemeyer Verlag.

Heisler, Ron. 2003. "Walking Stewart: A Forgotten Great Freethinker." Lecture to the Ethical Society. March 23, 2003. http://francisboutle.co.uk/pages.php?pID=69.

"Help of the Ghosts." 1854. *New-York Daily Tribune.* January 12, 1854: 4.

Hemans, Felicia. 1836. *The Poetical Works of Mrs. Felicia Hemans; Complete in One Volume.* Philadelphia: Grigg & Elliot.

Higginson, Thomas Wentworth. 1861. "Nat Turner's Insurrection." *Atlantic Monthly* 8 (46): 173–87.

Higginson, Thomas Wentworth. 1900. *The Writings of Thomas Wentworth Higginson: Army Life in a Black Regiment.* Cambridge, MA: Riverside Press.

Hoffmann, E. T. A. 2008. *Der Sandmann: Text und Kommentar.* Frankfurt am Main: Suhrkamp.

Holland, Henry. 1839. *Medical Notes and Reflections.* London: Longman, Omre, Brown, Green, and Longmsans.

Holmes, Oliver Wendell. 1859. "The Stereoscope and the Stereograph." *The Atlantic Monthly* 3 (20): 738–49.

Holmes, Oliver Wendell. 1868. "To Christian Gottfried Ehrenberg." *Littell's Living Age* 1279: 578.

Holmes, Oliver Wendell. 1887. *Selections from the Writings of Oliver Wendell Holmes: Arranged Under the Days of the Year.* Boston: Houghton Mifflin.

Holmes, Oliver Wendell. 1891. *The Autocrat of the Breakfast-Table: Every Man His Own Boswell.* Boston: Houghton, Mifflin.

Holmes, Oliver Wendell. 1891–95. *The Writings of Oliver Wendell Holmes.* 8 vols. Boston: Houghton, Mifflin and Company.

Homer. 2007. *The Odyssey.* Edited by Samuel Butler. Rockville, MD: Wildside Press.

"Horrible Accident." 1848. *Buffalo Weekly Republic.* October 3, 1848: 3.

244 REFERENCES

Horwitz, Tony. 2015. "Did Civil War Soldiers Have PTSD?" *Smithsonian Magazine*, January 2015. https://smithsonianmag.com/history/ptsd-civil-wars-hidden-legacy-180953652.

Howells, William Dean. 1887. "Editor's Study." *Harper's Magazine* 74: 482–86.

Hume, David. 1878. *A Treatise on Human Nature: Concerning Natural Religion: Being an Attempt to Introduce the Experimental Method of Reasoning into Moral Subjects; and Dialogues.* London: Longmans, Green.

Hunt, Sanford B. 1867. "The Negro as a Soldier: Reprinted from the Archives of the U.S. Sanitary Commission." *Quarterly Journal of Psychological Medicine and Medical Jurisprudence* 1 (2): 161–86.

Hunt, Sanford B. 1869. "The Negro as a Soldier." *Anthropological Review* 7 (24): 40–54.

Hustis, Harriet. 2003. "Deliberate Unknowing and Strategic Retelling: The Ravages of Cultural Desire in Charles Brockden Brown's Edgar Huntly." *Studies in American Fiction* 31 (1): 101–20.

Hutton, Fiona. 2015. *The Study of Anatomy in Britain, 1700–1900.* London: Routledge.

Hyrtl, Joseph. 1846. *Lehrbuch der Anatomie des Menschen.* Vienna: W. Braumüller.

The Illustrated Annuals of Phrenology and Physiognomy: For the Years 1865–1873. 1873. New York: Samuel R. Wells.

"Insurrection and Murder." 1831. *Raleigh Register.* September 1, 1831: 3.

Ione, Amy. 2016. *Art and the Brain: Plasticity, Embodiment, and the Unclosed Circle.* Leiden: Brill.

James, Henry. 1907. *Roderick Hudson.* New York: C. Scribner's Sons.

James, William. 1900. *On Some of Life's Ideals.* New York: Henry Holt.

James, William. 1981. *The Principles of Psychology.* 3 vols. Cambridge, MA: Harvard University Press.

Jameson, Fredric. 2015. *The Antinomies of Realism.* London: Verso.

Johnson, Walter. 1824. "Academy at Germantown, Aug. 12, 1842." Kislak Center for Special Collections, Rare Books and Manuscripts, University of Pennsylvania.

Jones, Ewan James. 2015. "John 'Walking' Stewart and the Ethics of Motion." *Romanticism* 21 (2): 119–31.

Jordan, John Emroy. 1962. *De Quincey to Wordsworth: A Biography of a Relationship.* Berkeley: University of California Press.

Kandel, Eric R. 2012. *The Age of Insight: The Quest to Understand the Unconscious in Art, Mind, and Brain, from Vienna 1900 to the Present.* New York: Random House.

Kaplan, Amy. 1988. *The Social Construction of American Realism.* Chicago: University of Chicago Press.

Karson, Michael. 2014. "We Shouldn't Treat Mental Illness Like Physical Illness." *Psychology Today.* https://psychologytoday.com/us/blog/feeling-our-way/201411/we-shouldn-t-treat-mental-illness-physical-illness.

Keats, John. 2009. *John Keats "Bright Star": Complete Poems and Selected Letters of John Keats.* London: Vintage Classic.

Kennedy, Alan. 1979. *Meaning and Signs in Fiction.* London: Palgrave Macmillan.

Kirkbride, Thomas Story. 1854. *On the Construction, Organization, and General Arrangements of Hospitals for the Insane.* Philadelphia: Lindsay & Blakiston.

Kittler, Friedrich A. 1993. *Draculas Vermächtnis: Technische Schriften.* Leipzig: Reclam.

Kittler, Friedrich A. 1999. *Gramophone, Film, Typewriter.* Stanford, CA: Stanford University Press.

Kittler, Friedrich A. 2009. *Musik und Mathematik: Hellas.* Munich: Fink.

REFERENCES   245

Kittler, Friedrich A. 2012. *Optical Media: Berlin Lectures 1999*. Cambridge, MA: Polity.

Kittler, Friedrich A. 2015. *Baggersee: Frühe Schriften aus dem Nachlass*. Edited by Tania Hron and Sandrina Khaled. Paderborn: Wilhelm Fink.

Kneeland, Samuel. 1851. "Report on Idiotic Crania, Idiocy, and Cretinism." *American Journal of the Medical Sciences* 21 (42): 349–63.

Knight, David. 2016. *Science in the Romantic Era*. New York: Routledge.

Knight, William Angus. 1889. *The Life of William Wordsworth*, vol. 1. Dover, NH: W. Paterson.

Knoper, Randall K. 2021. *Literary Neurophysiology: Memory, Race, Sex, and Representation in U.S. Writing, 1860–1914*. Oxford: Oxford University Press.

Knowlton, Charles. 1829. *Elements of Modern Materialism: Inculcating the Idea of a Future State in Which All Will Be More Happy Under Whatever Circumstances They May Be Placed than if They Experienced No Misery in This Life*. Adams, MA: Oakey.

Krämer, Sybille. "The Cultural Techniques of Time Axis Manipulation: On Friedrich Kittler's Conception of Media." *Theory, Culture and Society* 23 (7–8): 93–109.

Krämer, Sybille. 1994. "Geist Ohne Bewußtsein? Über Den Wandel in den Theorien vom Geist." In *Geist—Gehirn—Künstliche Intelligenz: Zeitgenössische Modelle Des Denkens*. Edited by Sybille Krämer, 88–112. Berlin: de Gruyter.

Krause, Sydney J. 1998. "Note on the Texts." In *Three Gothic Novels*. Edited by Sydney J. Krause, 908–9. New York: Library of America.

Kronshage, Eike. 2018. *Vision and Character: Physiognomics and the English Realist Novel*. New York: Routledge.

Lancaster, C. Sears. 1847. "Valentine: Or the Electric Telegraph, a Shocking Story." *The Court and Lady's Magazine* 30: 125–49.

Laster, Whitney. 2014. "Lamarckism Theory." In *Race and Racism in the United States: An Encyclopedia of the American Mosaic*. Edited by Charles A. Gallagher and Cameron D. Lippard, 682–84. Santa Barbara, CA: Greenwood.

Latour, Bruno. 2005. *Reassembling the Social: An Introduction to Actor-Network-Theory*. Clarendon Lectures in Management Studies. Oxford: Oxford University Press.

Laue, Max. 1895. *Christian Gottfried Ehrenberg: Ein Vertreter Deutscher Naturforschung im Neunzehnten Jahrhundert, 1795–1876*. Berlin: Julius Springer.

Laverty, Carroll Dee. 1951. "Science and Pseudo-Science in the Writings of Edgar Allan Poe." PhD dissertation, Duke University.

Lawrence, William. 1822. *Lectures on Physiology, Zoology, and the Natural History of Man: Delivered at the Royal College of Surgeons*. Salem, MA: Foote and Brown.

Leclerc, Georges-Louis, Comte de Buffon. 1792. *Buffon's Natural History: Containing a Theory of the Earth, a General History of Man, of the Brute Creation, and of Vegetables, Minerals, &c. From the French. With Notes by the Translator. In Ten Volumes*, vol. 5. London: J. S. Barr.

"Lectures." 1846. *Brooklyn Daily Eagle*, March 5, 1846: 2.

Leiter, Sharon. 2007. *Critical Companion to Emily Dickinson: A Literary Reference to Her Life and Work*. New York: Facts on File.

Leuschner, Pia-Elisabeth. 2000. *Orphic Song with Daedal Harmony: Die „Musik" in Texten der Englischen und Deutschen Romantik*. Würzburg: Königshausen & Neumann.

Levere, Trevor H. 1977. "Coleridge, Chemistry, and the Philosophy of Nature." *Studies in Romanticism* 16 (3): 349–79.

Levy, Eric P. 1996. "The Psychology of Loneliness in 'Wuthering Heights.'" *Studies in the Novel* 28 (2): 158–77.

246 REFERENCES

Liebig, Justus von. 1859. *Familiar Letters on Chemistry: In Its Relation to Physiology, Dietetics, Agriculture, Commerce and Political Economy.* Edited by John Blyth. London: Walton and Maberly.

Lind, Sidney E. 1947. "Poe and Mesmerism." *PMLA* 62 (4): 1077–94.

Loeffelholz, Mary. 1991. *Dickinson and the Boundaries of Feminist Theory.* Champaign: University of Illinois Press.

Lokhorst, Gert-Jan. 2006. "Descartes and the Pineal Gland." In *The Stanford Encyclopedia of Philosophy.* Edited by Edward N. Zalta. https://plato.stanford.edu.

Lovesay, Oliver. 1998. "The Other Woman in Daniel Deronda." *Studies in the Novel* 30 (4): 505–20.

Loving, Jerome M. 1982. *Emerson, Whitman, and the American Muse.* Chapel Hill: University of North Carolina Press.

Loving, Jerome M. 1999. *Walt Whitman: The Song of Himself.* Berkeley: University of California Press.

Lowenberg, Carlton. 1986. *Emily Dickinson's Textbooks.* Lafayette, CA: C. Lowenberg.

Ludwig, Sämi. 2002. *Pragmatist Realism: The Cognitive Paradigm in American Realist Texts.* Madison: University of Wisconsin Press.

Lukács, Georg. 1972. "Narrate or Describe?" In *Writer and Critic: And Other Essays.* Edited by Arthur Kahn, 110–48. London: Merlin Press.

Lumley, Benjamin. 1873. *Another World: Or, Fragments from the Star City of Montalluyah.* London: Samuel Tinsley.

Lux, Thomas. 1997. *New and Selected Poems, 1975–1995.* Boston, MA: Houghton Mifflin.

Lyons, William E. 1986. *The Disappearance of Introspection.* Cambridge, MA: MIT Press.

Mabbott, Thomas Ollive. 1978. "Mesmeric Revelation." In *The Collected Works of Edgar Allan Poe: Vol. III: Tales and Sketches.* Edited by Thomas O. Mabbott, 1024–28. Cambridge, MA: Belknap Press.

MacCabe, Colin. 1974. "Realism and the Cinema: Notes on Some Brechtian Theses." *Screen* 15 (2): 7–27.

MacFarquhar, Colin, and George Gleig, eds. 1797. *Encyclopædia Britannica: A Dictionary of Arts, Sciences, and Miscellaneous Literature,* vol. 3. Edinburgh: A. Well & C. MacFarquhar.

Macho, Thomas. 2013. "Second-Order Animals: Cultural Techniques of Identity and Identification." *Theory, Culture and Society* 30 (6): 30–47.

Mackey, Nathaniel. 2005. *Paracritical Hinge.* Contemporary North American Poetry. Madison, WI: University of Wisconsin Press.

Macmillan, Malcolm. 2002. *An Odd Kind of Fame: Stories of Phineas Gage.* Cambridge, MA: MIT Press.

Madden, William A. 1972. "Wuthering Heights: The Binding of Passion." *Nineteenth-Century Fiction* 27 (2): 127–54.

"The Magazines for August." 1884. *The Critic* 5 (31): 53–54.

"Magistrates of Irvine." 1816. *Caledonian Mercury.* October 12, 1816: 2.

"The Magnetic Telegraph—Some of Its Results: (From the Tribune)." 1845. *Littell's Living Age* 6 (63): 194–95.

Malane, Rachel Ann. 2005. *Sex in Mind: The Gendered Brain in Nineteenth-Century Literature and Mental.* New York: Peter Lang.

Man, Paul de. 1971. "The Rhetoric of Temporality." In *Blindness and Insight: Essays in the Rhetoric of Contemporary Criticism,* 187–228. Oxford: Oxford University Press.

REFERENCES   247

Marcus, Mordecai. 1962. "Melville's Bartleby as a Psychological Double." *College English* 23 (5): 365–68.

Marlow, James E. 1993. *Charles Dickens: The Uses of Time*. Selinsgrove, PA: Susquehanna University Press.

Martinez, Matias, and Michael Scheffel. 2016. *Einführung in Die Erzähltheorie*. Munich: C. H. Beck.

Masse, Michelle A. 2000. "'He's More Myself than I Am': Narcissism and Gender in Wuthering Heights." In *Psychoanalyses / Feminisms*. Edited by Peter L. Rudnytsky and Andrew Gordon, 135–53. Albany: State University of New York Press.

Matlak, Richard. 2002. "Swift's Aeolists and Coleridge's Eolian Harp." *Coleridge Bulletin* 20: 44–53.

Matoff, Susan. 2011. *Conflicted Life: William Jerdan, 1782–1869, London Editor, Author and Critic*. Sussex: Sussex Academic Press.

Mayo, Herbert. 1849. *Letters on the Truths Contained in Popular Superstitions*. Edinburgh: John David Sauerlænder, Messrs. Blackwood.

McCampbell, Jenna. 2017. "Macabre School Supplies: 19th-Century Dissection Sets." National Museum of American History, Smithsonian Institution. http://americanhist ory.si.edu/blog/19th-century-dissection-sets.

McCormack, Jerusha Hull. 2003. "Domesticating Delphi: Emily Dickinson and the Electro-Magnetic Telegraph." *American Quarterly*, 55 (4): 569–601.

McCuskey, Brian W. 1997. "'No Followers': The Victorian Servant Problem." *Nineteenth-Century Prose* 24 (1): 105–23.

McFarland, Thomas. 1969. *Coleridge and the Pantheist Tradition*. Oxford: Clarendon Press.

McGovern, Constance M. 1985. *Masters of Madness: Social Origins of the American Psychiatric Profession*. Hanover, NH: University Press of New England.

McLuhan, Marshall, Quentin Fiore, and Jerome Agel. 1967. *The Medium Is the Message*. Harmondsworth: Penguin.

McMaster, Juliet. 1987. *Dickens the Designer*. Basingstoke: Macmillan.

McMullen, Kevin, Jason Stacy and Stefan Schöberlein. "Walt Whitman at the *Aurora*: A Model for Journalistic Attribution." *Walt Whitman Quarterly Review* 37 (2019):107–115.

Melville, Herman. 1986. *Billy Budd and Other Stories*. Edited by Frederick Busch. London: Penguin Books.

Melville, Herman. 2011. *Moby-Dick*. San Francisco: Ignatius Press.

Mengham, Rod. 2001. *Charles Dickens*. Devon: Northcote House in association with the British Council.

Menke, Richard. 2008. *Telegraphic Realism: Victorian Fiction and Other Information Systems*. Stanford, CA: Stanford University Press.

Merritt, Keri Leigh. 2017. *Masterless Men: Poor Whites and Slavery in the Antebellum South*. Cambridge: Cambridge University Press.

Meyer, Susan. 1996. *Imperialism at Home: Race and Victorian Women's Fiction*. Ithaca, NY: Cornell University Press.

Miller, Craig W. 1964. "Coleridge's Concept of Nature." *Journal of the History of Ideas* 25 (1): 77–96.

Mills, Bruce. 2006. *Poe, Fuller, and the Mesmeric Arts: Transition States in the American Renaissance*. Columbia, NY: University of Missouri Press.

Milton, John. 2012. *Paradise Lost*. Edited by Dennis R. Danielson. London: Broadview Press.

248 REFERENCES

Minssen, Mins. 1997. "Die Windharfe Als Stimme Der Natur Zwischen Romantischer Tradition Und Technischer Vision." In *Äolsharfen: Der Wind Als Musikant*. Edited by Mins Minssen and Georg Krieger, 17–46. Frankfurt am Main: E. Bochinsky.

Mitchell, Domhnall. 2000. *Emily Dickinson: Monarch of Perception*. Amherst: University of Massachusetts Press.

Mitchell, Robert. 2013. *Experimental Life: Vitalism in Romantic Science and Literature*. Baltimore: Johns Hopkins University Press.

Mitchill, Samuel L., and Edward Miller. 1808. "[Review of Recherches Sur La Medeicine, Ou L' Application De La Chimie À La Médecine]." *The Medical Repository (and Review of American Publications on Medicine, Surgery and the Auxiliary of Science)* 4: 172–76.

Moglen, Helene. 1971. "The Double Vision of 'Wuthering Heights': A Clarifying View of Female Development." *Centennial Review* 15 (4): 391–405.

Monro, Henry. 1851. *Remarks on Insanity: Its Nature and Treatment*. London: John Churchill.

Montgomery, James. 1824. "The Effect of Music on Cain." In *Select British Poets, or, New Elegant Extracts from Chaucer to the Present Time*. Edited by William Hazlitt, 693–94. London: William C. Hall.

Moore, George. 1846. *The Use of the Body in Relation to the Mind*. London: Longman, Brown, Green, and Longmans.

Morgan, Nicholas. 1871. *Phrenology, and How to Use It in Analyzing Character*. London: Longmans, Green.

Moses, Wilson Jeremiah. 2004. *Creative Conflict in African American Thought: Frederick Douglass, Alexander Crummell, Booker T. Washington, W. E. B. Du Bois, and Marcus Garvey*. Cambridge: Cambridge University Press.

Moss, Kirby. 2003. *The Color of Class: Poor Whites and the Paradox of Privilege*. Philadelphia: University of Pennsylvania Press.

Muirhead, John H. 2013. *Coleridge as Philosopher*. London: Taylor & Francis.

Mullins, Maire. 2006. "Sexuality." In *A Companion to Walt Whitman*. Edited by Donald D. Kummings, 164–79. Malden, MA: Blackwell Publishing.

Nalbantian, Suzanne. 2004. *Memory in Literature: From Rousseau to Neuroscience*. Basingstoke: Palgrave Macmillan.

Napoleon the Third and His Court: By a Retired Diplomatist. 1865. London: J. Maxwell & Co.

"Negro Brains Small." 1906. *Fort Worth Star-Telegram*. August 26, 1906: 18.

"New Jersey News." 1879. *The Morning Post*. October 2, 1879: 1.

"New Publications." 1864. *Boston Daily Evening Transcript*. May 18, 1864: 4.

Newman, Beth. 1990. "'The Situation of the Looker-on': Gender, Narration, and Gaze in Wuthering Heights." *PMLA* 105 (5): 1029–41.

Nicholson, William. 1809. *The British Encyclopedia, or, Dictionary of Arts and Sciences*. 5 vols. London: Longman, Hurst, Rees, and Orme.

Nobele, E. de. 1835. "Observation de suicide." *Annales de Médecine Belge et Etrangère*, 1835.

Nodal, John Howard. 1877. "Coleridge in Manchester." *Notes and Queries* 7: 161–62.

Northrop, John Worrell. 1904. *Chronicles from the Diary of a War Prisoner in Andersonville and Other Military Prisons of the South in 1864: An Appendix Containing Statement of a Confederate Physician and Officer Relative to Prison Condition and Management*. Wichita, KS: Private Printing.

"Notices of New Books." 1846. *Brooklyn Daily Eagle*. November 16, 1846: 2.

"Nurture or Nature?" 1907. *Nashville Banner*. July 17, 1907: 4.

REFERENCES 249

Oerlemans, Onno. 2004. *Romanticism and the Materiality of Nature*. Toronto: University of Toronto Press.

"On Some New Contribution to the Chemical and Technical History of Coal Gas." 1869. *Journal of Gas Lighting, Water Supply and Sanitary Improvement* 18: 636.

Otis, Laura. 2011. *Networking: Communicating with Bodies and Machines in the Nineteenth Century*. Ann Arbor: University of Michigan Press.

"'Our New Dictionary' of Phrenology and Physiognomy." 1865. *American Phrenological Journal* (April): 108–12.

Packham, Catherine. 2012. *Eighteenth-Century Vitalism: Bodies, Culture, Politics*. London: Palgrave Macmillan UK.

Page, Norman. 1984. *A Dickens Companion*. London: Macmillan.

Pappas, Stephanie. 2014. "Conspiracy Theories Abound as U.S. Military Closes HAARP." NBC News. http:// nbcnews.com/science/weird-science/conspiracy-theories-abound-u-s-military-closes-haarp-n112576.

Parker, Reeve. 1972. "'Finer Distance': The Narrative Art of Wordsworth's 'The Wanderer.'" *ELH* 39 (1): 87–111.

Patents for Inventions: Abridgments of the Specifications Relating to Electrictiy and Magnetism. 1859. London: Great Seal Patent Office.

Patten, Robert L. 2017. *Charles Dickens and His Publishers*. Oxford: Oxford University Press.

Petri, Alexandra. 2016. "Ben Carson Reveals the Truth—There Are Two Donald Trumps." *Washington Post*, March 11, 2016. www.washingtonpost.com/blogs/compost/wp/2016/03/11/ben-carson-reveals-the-truth-there-are-two-donald-trumps.

Philips, Donna R. 2014. "Spinsterhood." In *All Things Dickinson: An Encyclopedia of Emily Dickinson's World*. Edited by Wendy Martin, 790–95. Santa Barbara, CA: Greenwood.

"Physiognomy." 1850. *New-York Daily Tribune*. December 5, 1850: 5.

Pierer, H. A., ed. 1828. *Encyclopaedisches Woerterbuch der Wissenschaften*, vol. 9. Alenburg: Literatur-Comptoir.

Pietikäinen, Petteri. 2007. *Neurosis and Modernity: The Age of Nervousness in Sweden*. Leiden: Brill.

Pinel, Philippe. 1806. *A Treatise on Insanity: In Which Are Contained the Principles of a New and More Practical Nosology of Maniacal Disorders Than Has Yet Been Offered to the Public*. Sheffield: W. Todd.

Pittman, John P. 2006. "Double Consciousness." In *The Stanford Encyclopedia of Philosophy*. Edited by Edward N. Zalta. https://plato.stanford.edu.

Poe, Edgar Allan. 1837. *The Philosophy of Animal Magnetism: Together with the System of Manipulating Adopted to Produce Ecstacy and Somnambulism the Effects and the Rationale*. Philadelphia: Merrihew & Gunn.

Poe, Edgar Allan. 1845. "Alfred Tennyson." *Broadway Journal* 2: 26.

Poe, Edgar Allan. 1850. "Poe on Headley and Channing." *Southern Literary Messenger*, 10: 608–12.

Poe, Edgar Allan. 1902. *The Complete Works of Edgar Allan Poe*, vol. 2. Edited by James A. Harrison. New York: T. Y. Crowell.

Poe, Edgar Allan. 1966. *The Letters of Edgar Allan Poe*, vol. 1. Edited by J. W. Ostrom. New York: Gordian Press.

Poe, Edgar Allan. 2006. *Fiction and Poetry: Complete and Unabridged*. New York: Barnes & Noble.

250 REFERENCES

Porden, Eleanor Anne. 1816. "The Skull and the Custom Officer." *The Gentleman's Magazine* 119 (2): 157–59.

Porter, Roy. 1995. "Shaping Psychiatric Knowledge: The Role of the Asylum." In *Medicine in the Enlightenment*. Edited by Roy Porter, 256–73. Amsterdam: Rodopi.

Poskett, James. 2014. "Skulls in Print: Scientific Racism in the Transatlantic World." https://cam.ac.uk/research/news/skulls-in-print-scientific-racism-in-the-transatlantic-world.

Poskett, James. 2020. *Materials of the Mind: Phrenology, Race, and the Global History of Science, 1815–1920*. Chicago: University of Chicago Press.

Posnock, Ross. 1997. "The Influence of William James on American Culture." In *The Cambridge Companion to William James*. Edited by Ruth A. Putnam, 322–42. Cambridge: Cambridge University Press.

Potter, Jonathan. 2016. "The Stereoscope and Popular Fiction: Imagination and Narrative in the Victorian Home." *Journal of Victorian Culture* 21 (3): 346–62.

Powell, J. H. 1993. *Bring Out Your Dead: The Great Plague of Yellow Fever in Philadelphia in 1793*. Philadelphia: University of Pennsylvania Press.

Powers, Samuel Ralph. 1920. *A History of the Teaching of Chemistry in the Secondary Schools of the United States Previous to 1850*. Minneapolis: University of Minnesota.

Preece, W. H. 1880. "The Telegraphic Achievements of Wheatstone." *Journal of the Royal Institution of Great Britain* 9: 297–304.

Prichard, J. C. 1831. "Soundness and Unsoundsness of Mind." In *The Cyclopaedia of Practical Medicine*. London: Sherwood, Gilbert, and Piper, and Baldwin and Craddock, vol. 4, 39–55.

Priestley, Joseph. 1782. *Disquisitions Relating to Matter and Spirit: To Which Is Added the History of the Philosophical Doctrine Concerning the Origin of the Soul, and the Nature of Matter: With Its Influence on Christianity. Especially with Respect to the Doctrine of the Pre-Existence of Christ*. Birmingham: J. Johnson.

"The Progress of the Electric Telegraph." 1860. *The Atlantic Monthly* 5 (29): 290–97.

Pullman, Bernard. 2001. *The Atom in the History of Human Thought*. Edited by Axel Reisinger. Oxford: Oxford University Press.

Raffa, Piero. 1967. *Avanguardia e Realismo*. Milano: Rizzoli.

Rampersad, Arnold. 1997. "Shadow and Veil: Melville and Modern Black Consciousness." In *Melville's Evermoving Dawn: Centennial Essayss*. Edited by John Bryant and Robert Milder, 162–80. Kent, OH: Kent State University Press.

Reed, Adolph L. 1997. *W. E. B. Du Bois and American Political Thought: Fabianism and the Color Line*. Oxford: Oxford University Press.

Reid, Thomas. 1785. *An Inquiry into the Human Mind: On the Principles of Common Sense*. Edinburgh: T. Cadell & J. Bell and W. Creech.

"Republican State Nominations." 1859. *Buffalo Morning Express*. October 24, 1859: 1.

"[Review of a New View of Insanity]." 1845. *British and Foreign Medical Review or Quarterly Journal of Practical Medicine and Surgery* 20 (1): 1–17.

"[Reviews]." 1884. *The Literary World* 30 (770): 76–77.

Reynolds, David S. 1995. *Walt Whitman's America: A Cultural Biography*. New York: Knopf.

Richardson, Alan. 2001. *British Romanticism and the Science of the Mind*. Cambridge: Cambridge University Press.

Richardson, Ruth. 2009. *Death, Dissection and the Destitute*. Chicago: University of Chicago Press.

REFERENCES 251

Rigby, Catherine E. 2004. *Topographies of the Sacred: The Poetics of Place in European Romanticism*. Charlottesville: University of Virginia Press.

Robinson, Kenneth. 1951. *Wilkie Collins: A Biography*. London: The Bodley Head.

Rogin, Michael Paul. 1985. *Subversive Genealogy: The Politics and Art of Herman Melville*. Berkeley: University of California Press.

Rohrbach, Augusta. 2002. *Truth Stranger than Fiction: Race, Realism, and the U.S. Literary Marketplace*. New York: Palgrave.

Ross, Charles Henry. 1869. "Changing Heads with Captain Copp: A Rattle-Brained Story." *Belgravia* 13: 106–12.

Ross, Charles Henry. 1870. "A Rattle-Brained Story." *The Spirit of Democracy*, January 25, 1870: 1.

Rothfield, Lawrence. 1992. *Vital Signs: Medical Realism in Nineteenth-Century Fiction*. Princeton, NJ: Princeton University Press.

Rothstein, William G. 1987. *American Medical Schools and the Practice of Medicine: A History*. Oxford: Oxford University Press.

Rowland, Ann Wierda. 2012. *Romanticism and Childhood: The Infantilization of British Literary Culture*. Cambridge: Cambridge University Press.

Rush, Benjamin. 1789. *Medical Inquiries and Observations*. Philadelphia: C. Dilly.

Rush, Benjamin. 1812. *Medical Inquiries and Observations upon the Diseases of the Mind*. Philadelphia: Kimber & Richardson.

Rush, Benjamin. 1948. *The Autobiography of Benjamin Rush: His "Travels Through Life" Together with His Commonplace Book for 1789–1813*. Edited by George W. Corner. Princeton, NJ: Princeton University Press.

Rush, Benjamin. 1981. *Benjamin Rush's Lectures on the Mind*. Edited by Eric T. Carlson, Jeffrey L. Wollock, and Patricia S. Noel. Worcester, MA: American Philosophical Society.

Ruston, Sharon. 2013. *Creating Romanticism: Case Studies in the Literature, Science and Medicine of the 1790s*. London: Palgrave Macmillan.

Sachs, Curt. 2006. *The History of Musical Instruments*. Mineola, NY: Dover Publications.

Sait, James E. 1974. "Tennyson, Mesmerism, and the Prince's 'Weird Seizures.'" *Yearbook of English Studies* 4: 203–11.

Sampson, George. 1979. *The Concise Cambridge History of English Literature*, vol. 3. Cambridge: Cambridge University Press.

Sanders, Andrew. 1994. "Introduction." In Charles Dickens, *Our Mutual Friend*. New York: Knopf Doubleday.

Sappol, Michael. 2002. *A Traffic of Dead Bodies: Anatomy and Embodied Social Identity in Nineteenth-Century America*. Princeton, NJ: Princeton University Press.

Schelling, Friedrich Wilhelm Joseph. 1858. *Friedrich Wilhelm Joseph Von Schellings Sämmtliche Werke*, vol. 3. Tübingen: J. G. Cotta.

Schlutz, Alexander. 2008. "Purloined Voices: Edgar Allan Poe Reading Samuel Taylor Coleridge." *Studies in Romanticism* 47 (2): 195–224.

Schöberlein, Stefan. 2012. "The Ever-Changing Nature of the Sea: Whitman's Absorption of Maximilian Schele de Vere." *Walt Whitman Quarterly Review* 30 (2): 57–77.

Schöberlein, Stefan. 2015. "Insane in the Membrane: Emily Dickinson Dissecting Brains." *Emily Dickinson Journal* 24 (2): 46–70.

Schöberlein, Stefan. 2016a. "Sheppard Lee." http://philadelphiaencyclopedia.org/archive/sheppard-lee/.

252 REFERENCES

Schöberlein, Stefan. 2016b. "Tapping the Wire: A Telegraphic Discourse." *American Literature* 88 (2): 269–300.

Schöberlein, Stefan. 2017. "Poe or Not Poe? A Stylometric Analysis of Edgar Allan Poe's Disputed Writings." *Digital Scholarship in the Humanities* 32 (3): 643–65.

Schöberlein, Stefan, and Stephanie M. Blalock. 2020. "'A Story of New York at the Present Time': The Historico-Literary Contexts of Jack Engle." *Walt Whitman Quarterly Review* 37 (3/4): 145–84.

Schwalm, Leslie A. 2018. "A Body of "Truly Scientific Work": The U.S. Sanitary Commission and the Elaboration of Race in the Civil War Era." *Journal of the Civil War Era* 8 (4): 647–76.

Schwartz, Howard. 2004. *Tree of Souls: The Mythology of Judaism*. New York: Oxford University Press.

Sconce, Jeffrey. 2000. *Haunted Media: Electronic Presence from Telegraphy to Television*. Durham, NC: Duke University Press.

Sen, Debashish. 2020. *Psychological Realism in 19th Century Fiction: Studies in Turgenev, Tolstoy, Eliot and Brontë*. Newcastle upon Tyne: Cambridge Scholars Publishing.

Seoane, M. 1854. *A Dictionary of the Spanish and English Languages*, vol 1. London: Longman, Brown, and Company.

Sforza. 1829. *The Vision of Noureddin: And Other Poems*. London: Hurst, Chance, & Co.

Shaffner, Taliaferro Preston. 1859. *The Telegraph Manual: A Complete History and Description of the Semaphoric, Electric and Magnetic Telegraphs of Europe, Asia, and America, Ancient and Modern*. New York: Pudney & Russell.

Shell, Marc. 2015. *Talking the Walk and Walking the Talk: A Rhetoric of Rhythm*. New York: Fordham University Press.

Shelley, Mary Wollstonecraft. 1823. *Frankenstein: Or, the Modern Prometheus*, vol. 1. London: G. and W. Whittaker.

Shelley, Percy Bysshe. 1905. *The Complete Works of Percy Bysshe Shelley: Miscellaneous Poems, 1817–1822*. London: Virtue.

Shermer, Michael. 2002. *The Skeptic Encyclopedia of Pseudoscience*, vol. 2. Santa Barbara, CA: ABC-CLIO.

Shook, John R. 2012. *Dictionary of Early American Philosophers*. New York: Continuum.

Shuttleworth, Sally. 2001. "Introduction." In George Eliot, *The Lifted Veil: And, Brother Jacob*. Edited by Sally Shuttleworth. London: Penguin.

Siegert, Bernhard. 2003. *Passage des Digitalen: Zeichenpraktiken der Neuzeitlichen Wissenschaften 1500–1900*. Berlin: Brinkmann & Bose.

Siegert, Bernhard. 2015. *Cultural Techniques: Grids, Filters, Doors, and Other Articulations of the Real*. New York: Fordham University Press.

Siemerling, Winfried. 2005. *The New North American Studies: Culture, Writing and the Politics of Re/Cognition*. London: Routledge.

Singer, Charles. 2011. *A Short History of Science to the Nineteenth Century*. Mineola, NY: Dover Publications.

Smith, C. U. M. 2007. "Brain and Mind in the 'Long' Eighteenth Century." In *Brain, Mind, and Medicine: Essays in Eighteenth-Century Neuroscience*. Edited by Harry A. Whitaker, C. U. M. Smith, and Stanley Finger, 15–28. New York: Springer.

Smith, Elihu Hubbard. 1973. *The Diary of Elihu Hubbard Smith, 1771–1798*. Edited by James E. Cronin. Philadelphia: American Philosophical Society.

Smith, Hamilton L. 1852. *Natural Philosophy for the Use of Schools and Academies: Illus. by Numerous Examples and Appropriate Diagrams*. New York: Newman and Ivison.

REFERENCES 253

Smith-Rosenberg, Carroll. 1993. "Subject Female: Authorizing American Identity." *American Literary History* 5 (3): 481–511.

Society for the Diffusion of Useful Knowledge. 1842. *The Biographical Dictionary of the Society for the Diffusion of Useful Knowledge*, vol. 1, pt. 2. London: Longman, Brown, Green, and Longmans.

Solly, Samuel. 1848. *The Human Brain: Its Structure, Physiology and Diseases*. Philadelphia: Lea and Blanchard.

Solomon, Eric. 1959. "The Incest Theme in *Wuthering Heights*." *Nineteenth-Century Fiction* 14 (1): 80–3.

"Some Biological Differences Noted." 1909. *Vardaman's Weekly* November 6, 1909: 2.

"Something About Physiology and Phrenology." 1847. *Brooklyn Daily Eagle*. March 7, 1847: 1.

Spradlin, Wilford W., and Patricia B. Porterfield. 1984. *The Search for Certainty*. New York: Springer-Verlag.

"Springfield and Vicinity." 1874. *Springfield Republican*. February 5, 1874: 1.

St. Armand, Barton Levi. 1984. *Emily Dickinson and Her Culture: The Soul's Society*. Cambridge: Cambridge University Press.

St. Dominique, C. de. 1874. *Animal Magnetism (Mesmerism) and Artificial Somnambulism: Being a Complete and Practical Treatise on That Science, and Its Application to Medical Purposes Followed by Observations on the Affinity Existing Between Magnetism and Spiritualism, Ancient and Modern*. London: Tinsley Brothers.

St. Evans, Jonathan B. T., and Keith Frankish. 2009. *In Two Minds: Dual Processes and Beyond*. Oxford: Oxford University Press.

Stamos, David N. 2017. *Edgar Allan Poe, Eureka, and Scientific Imagination*. Albany: State University of New York Press.

"'Standard'-Morality: Murder and Money." 1845. *Punch* 8: 162.

Stang, Richard. 1959. *The Theory of the Novel in England: 1850–1870*. New York: Columbia University Press.

Sterne, Jonathan. 2003. *The Audible Past: Cultural Origins of Sound Reproduction*. Durham, NC: Duke University Press.

Stewart, Dugald. 1792. *Elements of the Philosophy of the Human Mind*, vol. 1. Philadelphia: William Young.

Stewart, Garrett. 2015. *The Deed of Reading: Literature * Writing * Language * Philosophy*. Ithaca, NY: Cornell University Press.

Stewart, John "Walking." 1796. *Prospectus of a Series of Lectures: Or a New Practical System of Human Reason, Calculated to Discharge the Mind from a Great Mass of Error, and to Facilitate Its Labour in the Approximation of Moral Truth, Divested of All Metaphysical Perplexities and Nullities*. Philadelphia: Thomas Dobson.

Stewart, John "Walking." 1803. *Opus Maximum: Or, the Great Essay to Reduce the World from Contingency to System*. London: J. Ginger.

Stewart, John "Walking." 1810. *The Philosophy of Human Society: In Its Origin, Progress, Improvability, and Present Awful Crisis, Etc*. London: T. Egerton.

Stewart, John "Walking." 1837. *The Moral State of Nations: Or Travels over the Most Interesting Part of the Globe*. Middletown, NJ: G. H. Evans.

Stiles, Anne. 2006. "Robert Louis Stevenson's Jekyll and Hyde and the Double Brain." *SEL. Studies in English Literature 1500–1900* 46 (6): 879–900.

Stiles, Anne. 2019. "Brain Science." In *The Routledge Companion to Victorian Literature*. Edited by Dennis Denisoff and Talia Schaffer, 368–76. London: Routledge.

254 REFERENCES

Stiles, Anne. 2021. "'Dissecting Piece by Piece': Experimentalism in Late-Victorian Fiction." In *Literature and Medicine*. Edited by Andrew Mangham and Clark Lawlor, 38–55. Cambridge: Cambridge University Press.

Stiles, Anne, Stanley Finger, and François Boller, eds. 2013. *Literature, Neurology, and Neuroscience: Historical and Literary Connections*. Amsterdam: Elsevier.

Stiles, Ezra. 1933. *Letters and Papers of Ezra Stiles: President of Yale College, 1778–1795*. Edited by Isabel M. Calder. New Haven, CT: Yale University Library.

Stokes, Christopher. 2015. "Desacralizing the Sign: Tooke, Stewart and Romantic Materialism." In *Dynamics of Desacralization: Disenchanted Literary Talents*. Edited by Paola Partenza, 37–52. Göttingen: V&R Unipress.

Stolte, Tyson. 2022. *Dickens and Victorian Psychology: Introspection, First-Person Narration, and the Mind*. Oxford: Oxford University Press.

Stovall, Floyd. 1974. *The Foreground of Leaves of Grass*. Charlottesville: University Press of Virginia.

"A Strange Case." 1848. *Watertown Chronicle*. July 24, 1848: 1.

Strassler, Matt. 2012. "Matter and Energy: A False Dichotomy." https://profmattstrassler.com/articles-and-posts/particle-physics-basics/mass-energy-matter-etc/matter-and-energy-a-false-dichotomy/.

Sullivan, R. B. 1994. "Sanguine Practices: A Historical and Historiographic Reconsideration of Heroic Therapy in the Age of Rush." *Bulletin of the History of Medicine* 68 (2): 211–34.

Sullivan, Shannon, and Nancy Tuana. 2007. *Race and Epistemologies of Ignorance*. Albany: State University of New York Press.

Sully, James. 1881. "George Eliot's Art." *Mind* 6 (23): 378–94.

Tate, Gregory. 2009. "Tennyson and the Embodied Mind." *Victorian Poetry* 47 (1): 61–80.

Tate, Gregory. 2012. *The Poet's Mind: The Psychology of Victorian Poetry 1830–1870*. Oxford: Oxford University Press.

Tennyson, Alfred Tennyson. 1994. *The Works of Alfred Lord Tennyson*. Ware, MA: Wordsworth Classics.

Tennyson, Alfred Tennyson, Frederick Tennyson, and Charles Tennyson Turner. 1893. *Poems by Two Brothers*. London: Macmillan.

Thackeray, William Makepeace. 1946. *The Letters and Private Papers of William Makepeace Thackeray*, vol. 3. Edited by Gordon N. Ray. Cambridge, MA: Harvard University Press.

Thiher, Allen. 1999. *Revels in Madness: Insanity in Medicine and Literature*. Corporealities. Ann Arbor: University of Michigan Press.

Thompson, Charles Hutchinson. 1890. *A Genealogy of the Descendants of John Thomson of Plymouth, Mass. Also Sketches of Families of Allen, Cooke and Hutchinson*. Lansing, MI: D. D. Thorp.

Thompson, Lloyd G. 1884. "The Zig Zag Telegraph: A Tale of a Mind Diseased." *Atlantic Monthly* 54 (332): 184–90.

Thompson, Silvanus P. 1910. *The Life of William Thomson, Baron Kelvin of Largs*. London: Macmillan.

Thoreau, Henry David. 1906. *The Writings of Henry David Thoreau: Journal, 1837–1846*. Edited by Bradford Torrey. Boston: Houghton Mifflin.

"To Be Sold at Auction." 1860. *Bell's Life in London and Sporting Chronicle*. October 7, 1860: 1.

"Toussaint l'Ouverture and the Republic of Hayti." 1872. *Chambers's Miscellany of Instructive and Entertaining Tracts* 19 (151): 1–32.

REFERENCES 255

Transactions of the Medical Society of New Jersey: Centennial Meeting. 1866. New York: Daily Advertiser.

Traubel, Horace, ed. 1906. *With Walt Whitman in Camden*, vol. 1. Boston: Small, Maynard & Co.

Traubel, Horace. 1992. *With Walt Whitman in Camden*, vol. 7. Edited by Jeanne Chapman and Robert MacIsaac. Carbondale: Southern Illinois University Press.

Trippett, David. 2013. *Wagner's Melodies: Aesthetics and Materialism in German Musical Identity.* Cambridge: Cambridge University Press.

Trower, Shelley. 2009. "Nerves, Vibration and the Aeolian Harp." *Romanticism and Victorianism on the Net*, no. 54. https://doi.org/10.7202/038761ar.

Trower, Shelley. 2012. *Senses of Vibration: A History of the Pleasure and Pain of Sound.* New York: Continuum International Publishing.

Tumulty, Karen. 2020. "There Are Two Joe Bidens. The Wrong One Has Been Running for President." *Washington Post.* Accessed May 24, 2021. https://washingtonpost.com/video/editorial/opinion--there-are-two-joe-bidens-the-wrong-one-has-been-running-for-president/2020/02/11/53a4d81b-7521-4a5b-b4ea-bb1c413a3aed_video.html.

Turner, Nat, and Thomas R. Gray. 1831. *The Confessions of Nat Turner, Leader of the Late Insurrection in Southampton, VA.* Baltimore: Thomas R. Gray.

TV Tropes. "Amnesiac Hero." http://tvtropes.org/pmwiki/pmwiki.php/Main/AmnesiacHero.

Tyson, Lois. 1994. *Psychological Politics of the American Dream: The Commodification of Subjectivity in Twentieth-Century American Literature.* Columbus: Ohio State University Press.

Tytler, Graeme. 2012. "The Workings of Memory in Wuthering Heights." *Brontë Studies* 37 (1): 10–18.

Uno, Hiroko. 1998. "'Chemical Conviction': Dickinson, Hitchcock and the Poetry of Science." *Emily Dickinson Journal* 7 (2): 95–111.

Van Wyhe, John. 2004. *Phrenology and the Origins of Victorian Scientific Naturalism.* London: Routledge.

Vars, Fredrick E., and William G. Bowen. 1998. "Scholastic Aptitude Test Scores, Race, and Academic Performance in Selective Colleges and Universities." In *The Black-White Test Score Gap.* Edited by Christopher Jencks and Meredith Phillips, 457–79. Washington, DC: Brookings Institution Press.

Villanueva, Darío. 1997. *Theories of Literary Realism.* Albany: State University of New York Press.

"The Vision of Noureddin [Review]." 1829. *The Atheneaum*: 292–94.

Warren, Kenneth W. 1993. *Black and White Strangers: Race and American Literary Realism.* Chicago: University of Chicago Press.

Washington, Harriet A. 2008. *Medical Apartheid: The Dark History of Medical Experimentation on Black Americans from Colonial Times to the Present.* New York: Anchor.

Watson, Melvin R. 1949. "'Wuthering Heights' and the Critics." *Trollopian* 3 (4): 243–63.

Webster, Noah. 1828a. "Fancy." In *American Dictionary of the English Language.* Edited by Noah Webster. http://webstersdictionary1828.com/Dictionary/fancy.

Webster, Noah. 1828b. "Peninsula." In *American Dictionary of the English Language.* Edited by Noah Webster. http://webstersdictionary1828.com/Dictionary/fancy.

Webster, Noah. 1828c. "Signification." In *American Dictionary of the English Language.* Edited by Noah Webster. http://webstersdictionary1828.com/Dictionary/fancy.

256 REFERENCES

Weisbuch, Robert. 1998. "Prisming Dickinson: Or Gathering Paradise by Letting Go." In *The Emily Dickinson Handbook*. Edited by Gudrun Grabher, Roland Hagenbüchle, and Cristanne Miller, 197–224. Amherst: University of Massachusetts Press.

"What Are the Nerves?" 1862. *Cornhill Magazine* 5: 153–66.

White, Simon J., John Goodridge, and Bridget Keegan. 2006. *Robert Bloomfield: Lyric, Class, and the Romantic Canon*. Lewisburg, PA: Bucknell University Press.

Whitehead, Alfred North. 1953. *Science and the Modern World*. Cambridge: Cambridge University Press.

Whitman, Walt. 1855. *Leaves of Grass*. Brooklyn, NY: Andrew Rome. Whitman Archive.

Whitman, Walt. 1856. *Leaves of Grass*. Brooklyn: Fowler and Wells. Whitman Archive.

Whitman, Walt. c. 1859. "The Physique of the Brain." David M. Rubenstein Rare Book & Manuscript Library, Duke University.

Whitman, Walt. 1863. Letter of Walt Whitman to Louisa Van Velsor Whitman. June 22, 1863. Whitman Archive.

Whitman, Walt. 1892. *Leaves of Grass*. Philadelphia: David McKay. Whitman Archive.

Whitman, Walt. 2014. *Walt Whitman's Selected Journalism*. Edited by Douglas A. Noverr and Jason Stacy. Iowa City: University of Iowa Press.

Whitman, Walt. 2017. *Life and Adventures of Jack Engle*. Edited by Zachary Turpin. Iowa City: University of Iowa Press.

Whitman, Walter. 1844. "The Fireman's Dream: With the Story of a Strange Companion." *New York Sunday Times and Noah's Weekly Messenger*, March 31, 1844. Whitman Archive.

Whitman, Walter. 1846. "One Wicked Impulse! A Tale of a Murderer Escaped." *The Brooklyn Daily Eagle and Kings County Democrat*, September 7, 1846. Whitman Archive.

Whitman, Walter. 1849. "Letters from a Travelling Bachelor: Number III." *New York Sunday Dispatch*, October 28, 1849. Whitman Archive.

Whittier, John Greenleaf. 1894. *The Complete Poetical Works of John Greenleaf Whittier*. Edited by Horace E. Scudder. Boston: Houghton Mifflin.

"Wilkie Collins's Reading." 1874. *Springfield Republican*. February 6, 1874: 1.

Wickens, Andrew P. 2014. *A History of the Brain: From Stone Age Surgery to Modern Neuroscience*. London: Psychology Press.

Wigan, A. L. 1844. *A New View of Insanity: The Duality of the Mind Proved by the Structure, Functions, and Diseases of the Brain, and by the Phenomena of Mental Derangement, and Shewn to Be Essential to Moral Responsibility*. London: Longman, Brown, Green, and Longmans.

Williams, C. S. 1908. *Descendants of Captain Joseph Miller of West Springfield, Mass. 1698–1908*. New York: T. A. Wright.

Wilson, James Grant, and John Fiske, eds. 1888. *Appleton's Cyclopædia of American Biography*, vol. 2. New York: D. Appleton.

"Wind Instruments." 1881. *The Musical Herald* 2 (6): 121–22.

Winship, Michael. 2001. "Hawthorne and the 'Scribbling Women': Publishing the Scarlet Letter in the Nineteenth-Century United States." *Studies in American Fiction* 29 (1): 3–11.

Winslow, Forbes. 1856. "A Psychological Quarterly Retrospect." *Journal of Psychological Medicine* 9 (1): i–xxiv.

Wiskind-Elper, Ora. 2012. *Tradition and Fantasy in the Tales of Reb Nahman of Bratslav*. Albany: State University of New York Press.

Wolfe, Charles T. 2016. *Materialism: A Historico-Philosophical Introduction.* New York: Springer International Publishing.

"A 'Woman in White.'" 1868. *Boston Herald.* May 9, 1868: 1.

Woodman, William Bathurst, and Charles Meymott Tidy. 1877. *A Handy-Book of Forensic Medicine and Toxicology.* London: J. & A. Churchill.

Wordsworth, William. 1889. *The Complete Poetical Works of William Wordsworth.* Edited by John Morely. London: Macmillan.

Wright, Peter. 2005. "George Combe: Phrenologist, Philosopher, Psychologist (1788–1858)." *Cortex* 41 (4): 447–51.

Wrobel, Arthur. 1974. "Whitman and the Phrenologists: The Divine Body and the Sensuous Soul." *PMLA* 89 (1): 17–23.

Yolton, John W. 1984. *Thinking Matter: Materialism in Eighteenth-Century Britain.* Minneapolis: University of Minnesota Press.

Young, Robert M. 1970. *Mind, Brain, and Adaptation in the Nineteenth Century: Cerebral Localization and Its Biological Context from Gall to Ferrier.* Oxford: Clarendon Press.

Yuste, Rafael. 2015. "From the Neuron Doctrine to Neural Networks." *Nature Reviews. Neuroscience* 16 (8): 487–97.

Index

For the benefit of digital users, indexed terms that span two pages (e.g., 52–53) may, on occasion, appear on only one of those pages.

abolition, 134–35, 136–38, 141–42
Addison, Joseph, 136–37
Aeolian harp, 19–28, 29–37, 46–47, 54–56, 57–58, 204, 208–9
affect, 22–24, 96, 129–30, 147. *See also* emotions; feelings
agency, 42–43, 50, 58, 165–66
Agg, John, 24–25
Alderson, James and John, 24n.4
Aldini, Giovanni, 195–97
Alger, Horatio, 103
alienism. *See* psychiatry
All The Year Round, 169–70, 173
American Museum, The, 59–60
American Phrenological Journal, 69–70, 114n.19, 119
amnesia, 97
anatomy
    atomic theory and, 39, 54
    dissections and, 2–3, 29n.12, 100–2, 137–39, 141–42
    grandeur of, 1–2
    illustrations and, 3–6
    mind and, 16, 17–18, 74–75
    phrenology and, 100–6, 137–39, 141–42, 143, 160–61
    physiology and, 10–11, 24–25, 179–80
    poetry and, 1–2, 24–25, 26, 179–85
    proto-neuroscience and, 9–10
    racial science and, 137–39, 141–42, 143–44, 148
    rise of, 3
    See also brain, structures of; phrenology
Anderson, Charles R., 188–90

Anti-Insane Asylum Society, 169
apparat. *See* automaton
appearances, 44–46
apperception, 2–3, 145–48. *See also* Realism; senses
art, 19–25
artificial neural network, 227–28, 230. *See also* ChatGPT
Associationism, 28, 30–31, 37, 37n.25, 164–65, 210–11
Athanasius of Alexandria, 41n.34
Athenaeum, 76–78
Atlantic Monthly, 206–7, 219–20
Atlantic Telegraph Company, 217
atoms, atomic theory, 2–3, 37–55, 58, 71–72
Auerbach, Erich, 145–47
Aufschreibesystem (discourse network), 15–16, 96, 149
Austen, Jane, 98–63
automaton, 31–32, 39, 46–47
autopsies. *See* dissections

Babbage, Charles, 31–32
Bain, Alexander, 17–18, 212n.15
Barker, Fred, II, 102n.9
Barlow, John, 87–88
batteries, 2–3
Baumgartner, Barbara, 180
Beard, J. R., 131
Beck, Adolf, 219
Beecher, Henry Ward, 134
Begeisterung (enthusiasm), 36
behavior
    brain and, 3, 79, 100
    insanity and, 82

260 INDEX

behavior (*cont.*)
  phrenology and, 100, 105–6, 107–8,
    112–13, 124
Belgravia, "Changing Heads with Captain
    Copp," 157–60, 168
Berger, Hans, 215–16, 218–19
Berkeley, George, 28
Bersani, Leo, 147–48, 149
Bertillon, Alphonse, 144–45
Berzelius, Jöns Jacob, 51n.47
Bird, Robert Montgomery, 160–
    61, 191–94
  Sheppard Lee, 128–30, 162–68
Blanchet, François, 39
Bloomfield, Robert, 19–20, 31–32, 58
  Aeolian harp, 19–21, 23–24, 27n.9
  Farmer's Boy, The, 19–20
  Nature's Music, 20–21, 23–24
Bloomfield, Robert Henry, "The
    Mechanics of Music," 31–32,
    32n.15, 227–28
body, 14–15
  brain and, 161–62, 163–65, 179–83,
    194, 230–31
  identity and, 157–60, 164, 165–66
  mind and, 37, 56–57, 98, 161–62, 163–
    65, 230–31
  as model of perfectibility, 1–2
  as musical instrument, 23–25
  poetry and, 180–83, 190–92
  race and, 126–28
  self and, 157–60
  swapping, 157–60, 163–66
  See also brain; mind
Boos, Sonja, 6
Boshears, Rhonda, 102–3
brain, 2–5
  atomic theory and, 42–43, 46–49,
    51, 52–58
  behavior and, 3
  as bicameral coffin, 84–85
  body and, 161–62, 163–65, 179–83,
    194, 230–31
  as cave, 44, 46–48, 48n.42, 64–65
  consciousness and, 189–92
  deterministic accounts of, 161–
    62, 168
  discourse network of, 15–17

duality of, 3–4, 5–6, 8, 60–64, 74, 76–
    77, 78–79, 80–85, 89–93, 102, 110,
    164(*see also* Holland, Henry; Wigan,
    Arthur Ladbroke)
  "first century of," 9–11
  gendered differences in, 85–87
  harp of, 25–28, 29–35, 36–39, 52, 57–58,
    204, 208–9
  hemispheric, 75–76, 78–79, 80, 83–85
  human nature and, 1–2
  images of, 3–6
  injury to, 11, 94–100, 102–3, 111–13
  localization of mental functions, 6, 9–
    10, 11, 100–2
  as machine, 28, 29–37
  master-slave dialectic and, 86–87, 89–92
  as material system, 2–3, 5–6, 8–9, 10–
    11, 15–16, 18, 26–28, 29–37, 162–63,
    164–65, 180–94, 204, 228–32
  as medium, 15–16
  memory and, 164–65
  as mere mass, 2–4, 29–30, 31
  mind and, 10–11, 16–18, 29–37, 29n.12,
    42–43, 46–49, 51, 52–54, 58, 69, 73–
    75, 103, 161–62, 163–66, 168, 179–
    94, 205–6, 228–29, 230–31
  nervous system and, 28–31, 37n.25,
    205–7, 210–11, 225, 227–28
  personality and, 96–100
  physiology of, 179–91, 197–99
  popular discourse of, 6–8, 9–10, 18
  proto-neuroscience of, 9–10
  race and, 11, 87–92, 127–56
  in Romantic thought, 56–58
  self and, 11, 34, 73–75, 188–92
  senses and, 29n.13
  as ship, 88–89
  sleep and, 61–62 (*see also*
    somnambulism)
  society and, 124–25
  soul and, 25, 26–28, 29–30, 31, 32–34,
    183–86, 188–89
  as tabula rasa, 2, 31
  telegraph and, 198–225, 227–28
  thought and, 192–94
  as transmitter, 195–225, 227–28
  trauma and, 219–24
  unruly, 86–87

INDEX   261

See also body; mind
brain, structures of
  amygdala, 224
  Broca's area, 5
  cavum vergae, 5
  central sulcus, 5
  cerebellum, 5, 185
  cerebral cortex, 25–26n.7, 100
  cerebrum, 3–4, 46–47, 73–74, 80–81,
    84–85, 86–87, 93, 197
  cingulate cortex, 224
  cingulate gyrus, 5
  claustrum, 5
  corpus callosum, 25–26n.7, 180
  dura mater, 181*f*, 218–19
  fasciculus gracilis, 5
  fornix, 25
  geniculate (lateral, medial), 5
  globules, 197–98
  gray matter, 2–3, 5, 25–26n.7
  gyri, 180
  hippocampus, 25, 224
  hypothalamus, 25
  Lyra Davidis, 25, 58
  midbrain, 5
  pineal gland, 3–4, 25–26n.7
  pons, 25–26n.7
  prefrontal cortex, 96
  psalterium, 25, 58
  respiratory center, 5
  substantia nigra, 5
  sulci, 180
  white matter, 5
  See also hemispheres
Brande, William Thomas, 39–41, 49n.44
British and Foreign Medical Review, 78
British Medical Association, 218–19
British Spiritual Telegraph, The, 217
Broca, Paul, 144, 148, 156
  Broca's area, 5
Brock, William Hodson, 37n.25
Brontë, Charlotte, 192
Brontë, Emily, 8
  Wuthering Heights, 77–78, 79–88, 93
Brooklyn Daily Eagle, 114–15
Brooklyn Daily Times, 114n.19
Brown-Séquard, Charles-
  Édouard, 76–77

Brown, Charles Brockden
  Arthur Mervyn, 61n.4
  Edgar Huntly, 61–66, 69–70, 87–88, 93
  "Queries," 61–62
  Sky-Walk, 61
Brown, John, 161n.5
Brown, Thomas, 37n.24
Brown, William Wells, 132–35, 155
  Black Man, The, 132–33
  Clotelle, 133–34
  Negro in the American Rebellion, The, 136
  Rising Son, The, 132–33
Byron, George Gordon, Lord, 21

Cambridge Analytica, 18n.13
carceralism, 144
Carlyle, Thomas, 131, 132–33
Caton, Richard, 218–19
Chamber's Journal, 208
ChatGPT, 227–28, 230
chemistry, 37–44, 49, 50, 54. *See also*
  atoms, atomic theory
Christophe, Henri, 87–88
Civil War, 136–42, 149–50, 219–20
class
  phrenology and, 108–10, 113–14
  race and, 141
  realism and, 125
  transgressing, 172–73
Coffman, Peter, 137
cognition, 15n.7, 231–32
  atomic theory of, 39–40, 41–43, 52–
    54, 55–56
  embodied, 56–57
  material, 17–18, 36, 142, 192–94, 228–
    29, 230, 231–32
  neural theories of, 9–10
  race and, 129–30, 136–37, 142, 144,
    149, 155
  realism and, 150–51
  as reflexive, 26–28
  soul and, 26–28
  stereographs and, 149
  telegraphs and, 208, 214, 216, 219–20
  See also mind
Coleridge, Samuel Taylor, 8, 31, 32, 34–35,
    36–37, 46n.40, 49–50, 55–57
  "Dejection: An Ode," 32n.17

262 INDEX

Coleridge, Samuel Taylor (*cont.*)
 "Eolian Harp, The" 26–28, 30–31
 *Hints Towards the Formation of a
  More Comprehensive Theory of Life*,
  50, 71–72
 Theory of Life, 50–54
Coles, Abraham, 1–2, 3–4, 5–6, 16–17
 "Microcosm, The" 1–2, 16–17
Collins, Wilkie, 191–92
 American tour, 174
 Moonstone, The, 68–69n.14
 Woman in White, The, 168–77
Columbian Magazine, 69–70
Combe, George, 122, 123–24, 129–30,
 134–36, 137–38
 Constitution of Man, The, 103–6, 108–
  13, 134–35
 Essays on Phrenology, 68–
  69n.14, 162n.8
 System of Phrenology, 134–35
consciousness, 63–66, 67–69, 71, 74–
 80, 189–94
 "double consciousness," 92–93
Cooke and Wheatstone. *See* telegraph
Cornhill Magazine, 206–7
Coxe, John Redmond, 160–61
Cranch, Christopher Pearse, 215–16
 "Evening with the Telegraph-Wires,
  An," 208–9, 210–11
craniology, 2–3, 98–102
 racial science and, 143–45,
  148, 151–52
 See also phrenology
Cullen, William, 28
Cutter, Calvin, 179–80, 183–85
cybernetics, 14–15, 231
cyborgs, 208–14

Dalton, John, 37–39, 40–41, 42–43, 49
Darwin, Charles, 87–88n.44
Darwin, Erasmus, 56–57
David, King, 25
Davy, Humphrey, 37–39, 37n.25, 48–49
De Forest, John William, *Miss Ravenel*,
 149–52, 153–55
de Man, Paul, 35–36
de Quincey, Thomas, 40n.31
Deakin, Wayne George, 48–49

death, dying, 33n.19, 46–47, 68–72, 185,
 186–87, 190–91
 bicameral coffin, 84–85
 See also sleep
dementia, 231–32
Descartes, René, 3–4, 25–26n.7, 29–30
determinism, 34–35, 37–39, 54, 104–6,
 122, 161–62, 168
Dickens, Charles, 102–3, 104, 107, 123–24,
 169–70, 174
 Great Expectations, 97
 Our Mutual Friend, 107–13
Dickinson, Edward, 177–80
Dickinson, Emily, 8, 17, 23n.2, 174–77,
 179–80, 191–94
 "Brain, within it's Groove, The" 185–86
 "first Day's Night had come –, The"
  188–89
 "I felt a Cleaving in my Mind –," 185–86
 "I felt a Funeral, in my Brain,"
  180, 186–88
 "I've dropped my Brain – My Soul is
  numb," 185–86
 "Much Madness is divinest Sense –,"
  189–90
 "Spider sewed at Night, A" 180–
  83, 185–86
 "They put Us far apart –," 211–14
 "This Consciousness that is
  aware," 190–91
 "This is a Blossom of the Brain –,"
  183–85
discourse networks. See
 Aufschreibesystem
dissections, 2–3, 29n.12
 of Black bodies, 126–28, 131, 137–42
 of felons, 3, 195–96
 galvanism and, 195–96
 standardization of, 2–3
Dorsey, John Syng, 24n.5
Douglass, Frederick, 136
Downing, Harriet, "Remembrances of a
 Monthly Nurse," 97
du Bois-Reymond, Emil, 10–11
Du Bois, W. E. B., 92–93
dualism, 11, 18, 26–28, 52, 61–64, 74–85,
 89–93, 102, 110, 164, 229
 electromagnetism and, 69–70

of modern psychology, 17–18
resurgent, 14–15
dynamism, 49–50

Earle, Pliny, 178–79
Eclectic Review, 76–77
Edwards, Jonathan, 183–85
Ehrenberg, Christian Gottfried, 197–98,
204, 207
electroencephalography (EEG), 197,
216, 218–19
Eliot, George, 77–78, 102–3, 152, 155–56
Adam Bede, 97
Daniel Deronda, 153–55
Elliotson, John, 68–69n.14
embodiment. *See* body
Emerson, Ralph Waldo, 21, 55–56,
93n.51, 122–23
"Circles," 231–32
Emancipation Lecture, 132n.6
Representative Men, 132
emotions
madness and, 80–82
materiality of, 185–89, 190–91
phrenology and, 100–2
See also affect; feelings
empiricism, 10–11, 17–18, 145
encephalitis, 162–63. See also *yellow fever*
encephalography, 217
Encyclopædia Britannica, 29–30
eolian harp. *See* Aeolian harp
epilepsy, 34, 185–86
epistemology, 5–6, 14–15, 58,
142, 192–94
essence, 44–46, 48–49
ethnography, 143–52
Ethnological Society of London, 141–42

feelings
materiality of, 33–34, 36, 71
realism and, 145, 147–48
See also affect; emotions
Feltenstein, Rosalie, 88
Fern, Fanny, 169
Forbes, John, 78
Forstner, Freiherr von, 215–16
Foster, George, 195–96, 197
Foucault, Michel, 8–9, 144, 231

Fowler, Lorenzo N. and Orson, 107, 113–
17, 123–24, 125
Phrenology Proved, 116, 117–18
"Republicanism," 118–23
See also Whitman, Walt
Fraser's Magazine, 97
free will, 5–6, 11, 102–3, 104–6, 179–
80, 183–85, 188–89, 190–91. *See
also* agency
French Revolution, 41
function, 5–6, 75–76, 218–19, 224

Gage, Phineas, 94–97, 102
Galen, 2
Gall, Franz Joseph, 2–3, 10–11, 60, 98–64,
103–4, 105, 116–17, 122, 124, 161n.6
Galvani, Luigi, 195–96
galvanism, 195–97, 203–4
galvanometer, 217–19
Gaskell, Elizabeth, 78
Gauss, Carl Friedrich, 198–99
gender, 85–87
insanity and, 169–73, 174, 176–77
realism and, 147–48
transgressing, 172–73
Gilbert, Sandra M., and Susan Gubar, 85–
86, 85n.40, 171–73
Gilman, Sander, 155
Gilmore, Paul, 203–4
Google, 6
Gordon, Lyndall L., 185–86
Gorsky, Susan R., 80–81
Gould, Stephen Jay, 134n.10, 137–38, 143
Grant, A. Cameron, 104–5
grave robbing, 3
Gray, Thomas R., 127–28, 130–31
Great Man theory, 131–34, 135–36, 149
Greding, Johann Ernest, 29n.12
Grovier, Kelly, 40n.31
Guardian, 76–77

HAARP (High Frequency Active Auroral
Research Program), 58
Hagner, Michael, 3–4, 6, 62–63
Haitian Revolution, 129–30, 131. *See also*
L'Ouverture, Toussaint
Haller, John S., 137n.12
Hamilton, Cynthia. 134–35

264 INDEX

Hammond, William A., 141
Hanlon, Christopher, 231–32
Harlow, John, 94–96, 102
harmonics, 20–21, 31–32, 35–36
Harper's Weekly, 169–70, 173
Harrington, Anne, 8n.5
Hartley, David, 26–30, 32, 34–35, 37
  See also Associationism
Hawkins, John, 20
Hawthorne, Nathaniel, 104
Hayes, Albert H., 206n.9, *See also* "What
  Are the Nerves"
Hayles, N. Katherine, 14–15, 224–25
hearing. *See* listening; senses
Hegel, G. W. F., 48–49
Helmholtz, Hermann von, 8–9, 10–
  11, 15n.7
Hemans, Felicia, 24–25
hemispheres, 12, 60, 75–76, 78–79, 80, 83,
  87, 180, 228–29
  hemispheric lateralization, 9–10
  See also brain
Higginson, Thomas Wentworth, 127, 136–
  38, 174, 212
Hitchcock, Edward, 185–86
Holland, Henry, 92–93
  Medical Notes and Reflections, 75–77, 78
Holmes, Oliver Wendell, 76n.30, 145–47,
  192–94, 207
Horner, William Edmonds, 160–61
Hume, David, 28, 28n.11, 164–65
Hunt, Isaac H., 169
Hunt, James, 141–42, 144
Hunt, Sanford B., 137, 138–42, 143–
  44, 149–50
Hutton, James, 75–76
hypnotism, 69–70, 92–93. *See also*
  mesmerism
hypochondriasis, 80, 166–68
hysteria, 85

idealism, 40–41, 40n.29, 48–50, 51–53
*Illustrated Annuals of Phrenology and
  Physiognomy*, 107n.11
imagination, 20–21, 31, 56–57
imperialism, 144–45
insanity, 3, 59–61, 78–81, 82–85, 165–
  68, 177

gender and, 85–87
institutionalization and, 168–73,
  174, 176–79
irresistible impulse defense, 165–66
materiality of, 186–91
as metacognitive act, 11
moral insanity, 81–82
therapeutic programs for, 178–79
See also trauma; traumatic brain injury
intellect, 74, 81–82, 86, 110–12, 127, 129–
  30, 134–35, 136–37, 138–39, 142,
  143–44, 149–56
introspection, 2, 8–9, 10–11, 46–47
  scientific, 17–18

James, William, 10–11, 92–93, 124, 145–47
Jameson, Fredric, 147, 151–52

Kant, Immanuel, 15n.7
Kaplan, Amy, 103
Kennedy, Alan, 153
Kircher, Athanasius, 20–21
Kittler, Friedrich, 15–16, 20, 144–45, 231
Knoper, Randall, 143–44
Krämer, Sybille, 192–94

L'Ouverture, Toussaint, 131–32, 133–34,
  135, 137, 142
Lamarckism, 87–88, 105–6, 135–36
Lancaster, C. Sears, 199–203
Lathrop, Joseph, 59–60
Leicestershire Mercury, 77–78
Leuschner, Pia-Elisabeth, 34–35
Lever, Charles, 107n.11
Levere, Trevor H., 49–50
Lewis, Edmonia, 133–34
Liebig, Justus, 42n.35
Life Illustrated, 115–16. *See also* Fowler,
  Lorenzo N. and Orson
listening, 22–23, 203–4
literary studies, 8–9
literature, 8–9
  neurophysiology and, 7, 8–9
  photography and, 144
  relation to science, 2, 7, 11, 14–15, 56–
    57, 96–97, 124, 228–29
  telegraph and, 204–6, 208–14, 215–16
  See also poetry; Realism

INDEX 265

localization. *See under* brain
Locke, John, 25–26n.7
London Popular Record, 69–70
Lovelace, Ada, 31–32
Loving, Jerome, 114–15
Lukács, Georg, 144, 145–47
Lumley, Benjamin, 214–15
lunacy. *See* insanity
Lux, Thomas, 17n.11

Macho, Thomas, 231
madness. *See* insanity
Malane, Rachel Ann, 85n.41
materialism, 17–18, 30–31, 35–36, 48–
    49, 52n.51, 124, 134, 156, 160–61,
    161n.6, 162–63, 183–86
  religion and, 190–94
matter
  as active, 44–46, 50–54, 58
  atomic theory and, 37–43, 49–51, 53–
    56, 58, 71–72
  brain and, 2–3, 4–5n.2, 29–30, 164–65,
    183–85, 197–98, 228–32
  impenetrability of, 61–62, 62n.5, 71–72
  mechanized, 26–28, 31
  nature and, 44–56, 57, 58, 68–72
  as passive and inert, 31, 34–37, 49–51
  religion and, 4–5n.2, 36–39, 40n.29, 46,
    68, 71–72, 183–85, 203–4
  See also brain; mind
Mavrogenis, Spyridon, 7
Maxwell, James Clerk, 217–18n.18
McCabe, Colin, 147–48
McCormack, Jerusha, 212
McMaster, Juliet, 107n.12
mechanism, 39–40, 49–50, 54–55, 58
  determinism and, 37–39
media, mediation, 11, 198–225, 230–31
media theory, 8–9, 15–16
Medical and Surgical Reporter, 162–63
Medical Repository, 39
medical societies, 1–2
medicine, heroic, 161–63
  pre-professional history of, 9–10
melancholia, 22–24
Melville, Herman, "Benito Cereno," 88–
    93, 130–31
  Moby Dick, 88n.45, 192

memory, 28n.11, 29n.13, 44–46, 80–
    81, 164–65
Menke, Richard, 144, 153–54
mesmerism, 69–75, 197, 204n.8, 210–11.
    See also Poe, Edgar Allen
microscope, 2–3, 187–88, 197–98, 214–
    15, 224–25
Miller, Craig W., 50–51
Miller, Joseph, 59–60
mind, 2, 8–9
  agency and, 165–66, 168, 188–89
  atomic theory of, 39–49, 51, 52–56,
    58, 71–72
  body and, 161–62, 163–65, 230–31
  brain and, 5–6, 10–11, 16–18, 29–37,
    29n.12, 42–43, 46–49, 51, 52–54, 58,
    69, 73–75, 103, 161–62, 163–66, 168,
    179–94, 205–6, 228–29, 230–31
  cognition and, 231–32
  consciousness and, 63–64, 65–66, 67–
    69, 71, 74–76, 79–80, 189–90, 192–94
  as divided, 69
  duality of, 74–85, 89–93, 102
  embodiment and, 37, 98
  functional particularization of, 5
  imagination and, 20–21
  materiality of, 5, 7, 11, 51–56, 57,
    58, 61–62, 67–68, 70–75, 98, 102,
    161–62, 164–66, 168, 180–94, 203–
    4, 228–32
  mechanized, 29–39, 41
  metaphysics of, 124
  nature and, 44–49
  as passive, 57–58
  physiology of, 3–4, 102, 177
  poetry and, 44–49, 57
  politics and, 118–22
  race and, 127–56
  rationalist vision of, 28–29
  self and, 11, 159–60, 228–29, 231–32
  spirit and, 35–36
  subjectivity and, 194
  telegraph and, 199–225, 227–28
  thought and, 192–94
  See also brain; cognition;
    consciousness; self
Moglen, Helene, 85n.40
monism, 11, 14–15, 46, 69–70

266 INDEX

Montgomery, James, 24–25
moral insanity. *See under* insanity
Morse, Samuel. *See* telegraph
Morton, Samuel, 139–41, 142
motor function, 5
music, 31–32
myelinated axons. *See* nerves,
   nervous system

nature
   art and, 19–25
   cognition and, 35–37, 44–49, 57
   as dynamic, 49–58
   rationalist theory of, 49–50
   self and, 44–49
Naturphilosophie, 11, 51, 62–63
nerves, nervous system, 205–7, 210–11,
   225, 227–28
   brain and, 29–31, 37n.25
   duality of, 76–77
   ganglia, 74, 80n.37
   as hollow, 48–49n.43
   transmission and, 24–29, 37n.25
   See also "What Are the Nerves"
neurons, 37, 197–98, 205–6, 207,
   225, 227–28
neurophysiology, 5–6, 7, 8–9
neuroplasticity, 85–86
neuropsychology, 3
neuroscience, 2
   communication media and, 11
   early history of, 9–10, 15–17, 15n.7, 18
   modern cognitive, 18
New Jersey Medical Society, 1–2
New Materialism, 54n.53
New-York Daily Tribune, 215–17
New-York Magazine, The, 59–60
Newton, Isaac, 28–29
Nobele, E. de, 96–97
Northampton Lunatic Asylum, 177–79

Oerlemans, Onno, 40n.29
omoousia (one being), 41–42, 43–46
Opie, Amelia, 22–24
organicism, 48–49
Otis, Laura, 207

Packard, E. P. W. and Olsen, 169

particles. *See* atomic theory
Pease, Isaac T., 217
personality, personality changes, 96–100
Peyronie, François Gigot de la, 25–26n.7
philosophy, 5–6
photography, 144, 145. *See also* Realism
Phrenological Society of Philadelphia,
   160–61, 162–63
phrenology, 8, 9–10, 11, 18, 98–
   104, 114–16
   in America, 116–23
   belonging and, 107
   Black liberation and, 134–35
   class and, 108–10, 113–14
   free will and, 102–3, 104–6
   literary Realism and, 107–13, 123–25
   materialism and, 160–61
   monogenic theories of, 134–36
   morality and, 110
   politics and, 118–24, 125
   psychology and, 107, 124–25
   race and, 127–45
   same-sex attraction and, 113–14
   self-esteem and, 116–21
   society and, 124–25
Physick, Philip Syng, 24n.5, 160–
   61, 162–63
physics, 39–40, 49–51, 54–55
physiognomy, 18, 180–82. *See also*
   phrenology
physiology
   anatomy and, 10–11, 24–25, 179–80
   of atomic theory, 37–43
   brain and, 179–91, 197–99
   determinism and, 104–6
   free will and, 179–80
   gender and, 85–87
   mechanist understanding of, 54
   mind and, 32, 73–75, 102, 164–65, 177
   monism and, 69–70
   of nervous transmission, 24–29, 37n.25
pineal gland. *See under* brain,
   structures of
Pinel, Philippe, 7, 10–11
Poe, Edgar Allen, 165–66
   "Last Conversation of a Somnambule,
   The" ("Mesmeric Revelation"), 69–74
   Philosophy of Animal Magnetism, 73–75

INDEX 267

poetry, 1–2
    embodiment and, 180–83, 190–92
    mechanical, 31–33, 34–35
    mind and, 44–49, 57
    origins of, 20, 26
    science and, 186–88
    See also literature
Porden, Eleanor Anne, 25–26n.7
Porterfield, Patricia B., 204
posthumanism, 14–15, 231
postmodernism, 192–94
Prichard, James Cowles, 81–82
Priestley, Joseph, 28, 30–31, 37, 61–62, 161–62
psyche, 5–6, 8–9, 17–18, 23–24, 60–
    61, 77–79
psychiatry, 7
    clinical, 3
    material nature of mental illnesses,
        18, 29n.12
    therapeutic approaches, 178–79·
psychoanalysis, 7, 8–9
psychograph, 215–17
psychology, 5–6, 8–9, 16–18, 229
    belonging and, 107
    character and, 88–92
    comparative, 3
    empirical, 10–11
    everyday, 124–25
    intellect and, 124, 143–45, 152, 153–56
    mechanist, 54
    phlebotomizing, 162–63
    phrenology and, 124–25
    race and, 153–56
    social, 124–25
psychometrics, 11, 143–56. *See also* race
PTSD (post-traumatic stress
    disorder), 219–24
Punch, 77–78

race
    anti-Blackness and, 128–29
    Black genius and, 131–34
    brain and, 11, 87–92
    dissections of Black bodies, 126–28,
        131, 137–42
    phrenology and, 127–45, 153–56
    psychometrics and, 143–56
    racial hierarchism, 143–56

racial mixing, 130
    Realism and, 144–56
    scientific racism, 134–35, 143–52
    white supremacy and, 131–34, 141–42
Raffa, Piero, 123–24
Raleigh Register, 128n.4
Ramón y Cajal, Santiago, 198, 225
ratiocination, 43–44, 47–49
rationality, 5–6, 11, 28–29
    discord and, 86
    identity and, 166–68
    mechanism and, 39–40, 49–50
    physiological basis of, 102
    subjectivity and, 231
Realism, 102–3
    experience and, 145–48
    phrenology and, 107–13, 123–25
    psychology and, 143–56
    race and, 144–56
    See also literature
reason. *See* rationality
Reconstruction, 144–45, 147–48, 149, 152
Redfield, James W., 76–77, 87
Reid, Thomas, 30–31
religion, 41n.34
    Aeolian harp and, 22–24, 25, 26–28,
        31, 32–37
    mind-materialism and, 2, 4–5n.2, 25,
        25–26n.7, 26–28, 29–31, 32–37,
        40n.29, 44–46, 62n.5, 68–69, 70–72,
        98, 183–86, 190–94, 203–4
    phrenology and, 112–13, 115n.22,
        121n.31, 129–30
    race and, 129–30, 138n.13
    self and, 188–89
    See also soul
Remak, Robert, 207
Republicanism, 118–20
Richardson, Alan, 56–57, 98
Richerand, Anthelme, 25–26n.7
Rigby, Catherine, 44
Rogin, Michael Paul, 88
Romanticism, 19–20
    Aeolian harp and, 19–28, 46–47, 54–56,
        57–58, 68n.13, 204
    atomic theory and, 40–41, 44–56, 58
    brain and, 31–37, 56–58
    chemistry and, 37–39

268 INDEX

Romanticism (*cont.*)
electricity and, 203–4
identity and, 159–60
imagination and, 20–21, 31
mind-materialism and, 11
self and, 44–49
vitalism and, 52n.51
Rosecrans, William Starke, 221–22
Rothfield, Lawrence, 123–24
Royal College of Surgeons, 195–96, 208
Rush, Benjamin, 9–11, 28, 29–30, 59–63,
73–74, 93, 160–63, 192–94
*Medical Inquiries and
Observations*, 166–68
Russell, Ira, 137–41, 142, 143–44

Sanders, Andrew, 107
Sanderson, Jeremiah B., 133–34
Sanitary Commission, 137,
137n.12, 138–42
scale, 37–39
Schädellehre. See craniology; phrenology
Schelling, Friedrich Wilhelm Joseph, 37–
39, 50–51
Scholastic Aptitude Test (SAT), 155–56
Schwalm, Leslie A., 137n.12, 141
Schwann cells, 5
science
culture and, 15–16
literature and, 11, 14–15
religion and, 41n.34
transatlantic circulation of, 7
See also religion
seeing, sight, 145–48. *See also* senses;
vision, visual
Seelenschreiber (soul writer). *See*
psychograph
self
atomic theory of, 41–49
body and, 157–60, 164, 165–66
character and, 88–92
cognition and, 26–28
death and, 67–69, 70–71, 73–74
dual, 74–75, 79–80
gender and, 85–87
identity and, 5–6, 59–61, 64–65, 67–69,
157–60, 164, 165–68, 188, 191
(im)mortality of, 6

mind and, 11, 159–60, 168, 228–
29, 231–32
nature and, 44–49
race and, 87–92
Romanticism and, 44–49
soul and, 188–89
self-control, 5–6
self-esteem, 116–21
Sensationalism, 28, 30–31, 32, 37, 39–
40, 50, 57
senses, 2n.1
cognition and, 28–29
deprivation of, 44, 46–48
insanity and, 189–90
rationalism and, 28–29
See also hearing; seeing, sight;
vision, visual
sensorium commune, 29–30
Sforza, "The Spectre Harp," 32–34
Shadd, Mary Ann, 133–34
Shelley, Mary, *Frankenstein*, 195–96
Shelley, Percy Bysshe, "Ode to the West
Wind," 21, 34, 54–55
Shuttleworth, Sally, 8–9
Siegert, Bernhard, 15n.7, 231
slave revolts, 87–88, 89–92, 126–31
slavery, 3, 121–22
sleep, 60–62, 63–64, 65, 68–69. *See also*
somnambulism
Smith, C. U. M., 8n.5
Smith, Elihu Hubbard, 61
Smith, Lydia, 169
Smith, William Benjamin, 142
sociopathy, 81–82
solar plexus, 74–75
Sömmerring, Samuel von, 7
somnambulism (sleepwalking), 60–76,
87–88, 92–93. *See also* sleep
soul, 2, 35–36, 44–46, 62n.5, 71, 104–
5, 203–4
brain and, 4, 4–5n.2, 25, 25–26n.7, 29–
34, 183–86, 188–89, 192–94
cognition and, 26–28
self and, 188–89
See also Brown, Charles Brockden;
Poe, Edgar Allen; religion;
somnambulism; Tennyson Alfred
Spencer, Herbert, 67–68

INDEX 269

spine, spinal cord, 5, 76–77, 97–98n.5, 185.
  See also nerves, nervous system
spirit. *See* soul
Spiritual Telegraph, The, 217
Spiritualism, 197, 203–4, 217
Spradlin, Wilford, 204
Springfield Republican, 174
Spurzheim, Johann Gaspar, 2–3, 10–11,
  98–64, 103–4, 105, 114–15, 116–
  17, 122
St. Armand, Barton Levi, 180–82
Stacy, Jason, 114n.20
Stang, Richard, 106
stereograph, 145–55, 156
  See also Broca, Paul
stereoscope, 2–3
Sterne, Jonathan, 230–31
Stevenson, Robert Louis, 77–78
Stewart, John "Walking," 40–41, 44–46, 49
  Apocalypse of Nature, The, 41n.32
  Opus Maximum, 40n.31, 41–43
  Philosophy of Human Society, The, 41
Stiles, Anne, 8–10
Stiles, Ezra, 59–60
Stole, Tyson, 229
Stowe, Harriet Beecher, 104
  Uncle Tom's Cabin, 130
subjectivity
  materialism and, 32–37
  mind and, 165–66, 194
  neurophysiology and, 8–9, 14–15
  rationality and, 231
  See also self
sublime, 19–20
suicide, 65–68, 69
Sully, James, 152
systems, 41–43

tabula rasa, 2–3, 28, 31
Tate, Gregory, 75n.29
technology, 19–24. *See also* Aeolian harp;
  automaton; telegraph
telegraph, 2–3, 198–225, 227–28
Tennyson, Alfred, 192
  "Two Voices, The," 67–71, 93
Thackeray, William Makepeace, 137n.11
thinking machines, *See* Lovelace, Ada
Thompson, James, 22–24

Thompson, Lloyd G., 219–24
Thomson, William, 217–19
Thoreau, Henry David, 21, 203–4
thought. *See* cognition
Todd, Mabel Loomis, 174–75
trance, 92–93. *See also* mesmerism
transatlantic cables, 204–5, 207–9. *See also*
  telegraph
transcription machines, 215–16
translation, 22–23
transmission, 21, 23–29, 37n.25
Traubel, Horace, 114–15, 125
trauma, *See* PTSD; traumatic brain injury
traumatic brain injury, 97–98, 111–13
Trippett, David, 215–16
Trower, Shelley, 27n.8, 56–57
Turing, Alan, 192–94
Turner, Nat, 126–31, 133–34, 135, 136, 142
Twain, Mark, 174

universalism, 37–39, 42–43, 48–49
Uno, Hiroko, 186–88
Ure, Andrew, 195–96n.1
utopianism, 122

vibrantiuncles, 37. *See also* Hartley, David
vibrations, 41n.32
vision, visual, 2–3, 37n.25, 82, 144, 145–
  47, 209–10. *See also* senses
vitalism, 51–52, 69–70, 197, 203–4

Wagner, Adolphus Theodore, 215–
  17, 218–19
Warren, Kenneth W., 145, 147–48, 152
Weber, Wilhelm Eduard, 198–99
Weisbuch, Robert, 192–94
Westmorland Gazette, The, 77–78
"What Are the Nerves," 206–7. *See also*
  nerves, nervous system
Wheatly, Phillis, 133–34
Whitaker, Harry, 8n.5, 102–3
Whitehead, Alfred, 5–6, 17–18
Whitehouse, Wildman, 208
Whitman, Walt, 8, 41–42, 102–3, 123–24,
  125, 141
  "Fireman's Dream, The" 97
  Leaves of Grass, 113–14, 118–19, 122–
  23, 192

270 INDEX

Whitman, Walt (*cont.*)
  "Letters from a Travelling Bachelor,"
    113n.17
  Life and Adventures of Jack
    Engle, 113–14
  "One Wicked Impulse!," 165n.12
  on phrenology, 114–16, 118–20, 122
  "Poem of Many in One" ("By Blue
    Ontario's Shore"), 118–22
  "Poem of Salutation," 209–10
  postmortem, 17
  "This Compost," 42n.35
  See also Fowler, Lorenzo N.
    and Orson
Whittier, John Greenleaf, 174, 204–6

Wigan, Arthur Ladbroke, *A New View of*
  Insanity, 9–10, 77–88, 92–93
Willis, Thomas, 28
Wilson, William J., 133–34
wind harps. *See* Aeolian harp
Wordsworth, William, 21, 44, 53–
  54, 56–57
  "Cave of Staffa," 48n.42
  Excursion, The, 28–48
  Prelude, The, 44–47
  "Wanderer, The" 47–49
Wundt, Wilhelm, 10–11, 17–18

yellow fever, 61, 161, 162–63
Young, Robert, 5–6